Springer Remote Sensing/Photogrammetry

More information about this series at http://www.springer.com/series/10182

Alberto Refice • Annarita D'Addabbo
Domenico Capolongo
Editors

Flood Monitoring through Remote Sensing

Springer

Editors
Alberto Refice
Istituto di Studi sui Sistemi
Intelligenti per l'Automazione (ISSIA)
Consiglio Nazionale delle Ricerche (CNR)
Bari, Italy

Annarita D'Addabbo
Istituto di Studi sui Sistemi
Intelligenti per l'Automazione (ISSIA)
Consiglio Nazionale delle Ricerche (CNR)
Bari, Italy

Domenico Capolongo
Department of Earth and
Environmental Sciences
University of Bari
Bari, Italy

ISSN 2198-0721 ISSN 2198-073X (electronic)
Springer Remote Sensing/Photogrammetry
ISBN 978-3-319-63958-1 ISBN 978-3-319-63959-8 (eBook)
https://doi.org/10.1007/978-3-319-63959-8

Library of Congress Control Number: 2017955831

Printed on acid-free paper

This Springer imprint is published by Springer Nature
The registered company is Springer International Publishing AG
The registered company address is: Gewerbestrasse 11, 6330 Cham, Switzerland

Preface

When we started our involvement in flood monitoring a few years ago, we were aware of the large number of researchers coming from different fields who contribute to several different aspects of producing and validating high-resolution flood maps from remotely sensed and other data. Each of the relevant research fields, including image and data processing, computing, hydrology, and geomorphology, builds upon a mature and robust tradition.

Nevertheless, the steadily increasing flow of high-resolution, open-access remote sensing data is leading to a paradigm change in the field of flood and, more generally, hazard monitoring.

The availability of high-resolution images of large portions of the Earth's surface in several electromagnetic bands – on a regular, high-frequency basis – is triggering the development of new algorithms and technologies to extract information and knowledge about subtle effects and temporal developments of inundation events. This in turn requires a synthesis work from traditionally separate research communities to develop common languages, standards, and methods.

This book presents a snapshot of some of the most up-to-date examples of approaches to such a daunting endeavor. By reading the various chapters from the outstanding research groups who agreed to contribute, whom we warmly thank, the reader will gain an understanding of the wide spectrum of skills and background necessary to extract meaningful flood information from remotely sensed and ancillary data. More importantly, we hope this book will serve as a reference toward the long-term objectives mentioned above, with the final goal of increasing awareness and easing the impact of such catastrophic events on people.

Bari, Italy
July 2017

Alberto Refice
Annarita D'Addabbo
Domenico Capolongo

Contents

Methods, Techniques and Sensors for Precision Flood Monitoring Through Remote Sensing ... 1
Alberto Refice, Annarita D'Addabbo, and Domenico Capolongo

Remote Sensing as a Tool for Analysing Channel Dynamics and Geomorphic Effects of Floods ... 27
Margherita Righini and Nicola Surian

The Use of DEM-Based Approaches to Derive a Priori Information on Flood-Prone Areas ... 61
Salvatore Manfreda, Caterina Samela, and Tara J. Troy

River Flood Forecasting System: An Interdisciplinary Approach ... 81
Viacheslav Zelentsov, Ilya Pimanov, Semen Potryasaev, Boris Sokolov, Sergey Cherkas, Andrey Alabyan, Vitaly Belikov, and Inna Krylenko

Monitoring Flood Extent and Area Through Multisensor, Multi-temporal Remote Sensing: the Strymonas (Greece) River Flood ... 101
Alberto Refice, Annarita D'Addabbo, Francesco Paolo Lovergine, Khalid Tijani, Alberto Morea, Raffaele Nutricato, Fabio Bovenga, and Davide Oscar Nitti

Adaptive SAR Image Processing Techniques to Support Flood Monitoring from Earth Observation Data ... 115
Silvana G. Dellepiane and Laura Gemme

Flood Mapping in Vegetated and Urban Areas and Other Challenges: Models and Methods ... 135
Nazzareno Pierdicca, Luca Pulvirenti, and Marco Chini

Data Fusion Through Bayesian Methods for Flood Monitoring from Remotely Sensed Data 181
Annarita D'Addabbo, Alberto Refice, Domenico Capolongo, Guido Pasquariello, and Salvatore Manfreda

Erratum E1

Index 209

The original version of this book was revised. An erratum to this book can be found at https://doi.org/10.1007/978-3-319-63959-8_9

Methods, Techniques and Sensors for Precision Flood Monitoring Through Remote Sensing

Alberto Refice, Annarita D'Addabbo, and Domenico Capolongo

Abstract In this chapter we introduce the subject matter of the book, highlighting the many aspects of signal and image processing, modelling and interpretation involved in the production of high-precision flood maps from remotely sensed data. For each field linked to the flood mapping activity, we summarize some basic definitions and research aspects, reviewing briefly some of the most recent applications and citing appropriate literature. We then introduce organically the subjects treated in each of the following chapters.

Keywords Flood monitoring • Optical and SAR remotely sensed data • Automatic flood detection techniques • Geomorphological mapping

1 Introduction

Climate change and anthropogenic pressure are increasing frequency and intensity of flood phenomena. In many countries, this causes increasing damages and threats to life, property and cultural heritage. Countermeasures and preparedness directives are being taken at local up to continental level [1, 71, 88]. Several environmental agencies around the world support communities in protecting lives and property, by providing tools and information to help them understand local flood risk and make cost-effective mitigation decisions. Such information is usually provided in the form of maps. In particular, flood inundation maps are used mainly for:

- Forecast scenarios
- Mitigation and planning – flood risk analyses
- Timely response

A. Refice (✉) • A. D'Addabbo
ISSIA-CNR, Bari, Italy
e-mail: refice@ba.issia.cnr.it; daddabbo@ba.issia.cnr.it

D. Capolongo
Department of Earth and Environmental Science, University of Bari, Bari, Italy
e-mail: domenico.capolongo@uniba.it

A. Refice et al. (eds.), *Flood Monitoring through Remote Sensing*,
Springer Remote Sensing/Photogrammetry,
https://doi.org/10.1007/978-3-319-63959-8_1

- Damage assessment
- Environmental and ecological assessments

These maps are traditionally created using hydraulic and topographic modelling, to simulate and visualize a wide range of flooding scenarios. However, the methods for estimation of flood risk maps based on inundation models and statistics are sometimes not very accurate, because of the large uncertainties in land surface topography, hydrological modelling and statistics [4]. Furthermore, the controlling risk factors change geographically and temporally, and an understanding of their physical and spatiotemporal characteristics is essential for the projection of future flood risks and the development of effective flood mitigation measures [109].

Remote sensing plays an important role in all the phases of hazard management, from the alert and preparedness phase to the emergency management and civil protection and up to damage assessment for risk reduction.

Research and operational activities in the field of flood monitoring and prevention benefit from availability of both optical and radar remotely sensed (RS) data. RS data offer some long recognized advantages, such as the capability of providing synoptic information over wide areas at low costs, the reliability of data acquisition schedule, the immunity to local hindrances such as site accessibility or dangerous environmental conditions, etc. RS data availability has increased over the past few years, thanks to renewed efforts by several national space agencies in developing and operating remote sensors with characteristics 1–2 orders of magnitude better than those related to the legacy generation. As a result, such new-generation data give more opportunities of use and allow new applications, often opening new research fields. In the case of flood monitoring, high-resolution, multi-temporal data give the chance to improve significantly the performances of forecast, alert and post-event monitoring of inundation events. However, as is often the case, new opportunities and applications pose new challenges: the nature of the data, the different size and scale of the objects and the processes which can be now investigated, even the sheer quantity of available data, all require new or more powerful tools to be suitably dealt with. In the particular field which is the subject of this book, devising reliable and robust methods to exploit remotely sensed data to their full potential involves recurring to a large spectrum of fields of expertise and knowledge, which are not all usually available to single scientists or research groups.

In the present introduction, we propose a short review of the main aspects linked to last-generation remote sensing data analysis as applied to flood monitoring, by briefly illustrating the links to research fields and theoretical backgrounds as can be found at the present time.

2 Overview of Flood Monitoring Systems and Remote Sensing Approaches

Floods are broadly defined as events in which water covers land areas which are not normally covered by water [28]. This definition covers a wide variety of events under the geomorphic framework, as regards size, duration or cause. Consider, for instance, the extremes of recurrent floods caused by large rivers, which can cover thousands of hectares, and flash floods, typically affecting small areas in a short time, but with potentially destructive effects.

Linked to the nomenclature connected to flood events is their management, which includes monitoring as a measure to reduce damages connected to events. Flood monitoring activities can be roughly divided into three broad sets, according to the stage of operations with respect to the event occurrence [95]:

1. Forecasts
2. Emergency monitoring
3. Damage assessment

In spite of this perhaps somewhat artificial separation in time, the three kinds of activity are actually closely intertwined. For instance, effective forecasts help directing emergency monitoring efforts to the most affected areas and infrastructures. Also, precise damage assessment can be used as benchmark to improve the forecast capabilities of current systems. The above subdivision determines the type of RS approaches and data used in each activity. Forecasts involve mainly meteorological information, so low- to medium-resolution optical data are usually involved. In emergency monitoring, the emphasis is clearly on fast response and relatively high resolution, so a potentially wide variety of sensors can be involved, working in both the optical and microwave spectral regions. However, the typical bad weather conditions associated to flood events, with thick cloud coverage present at least during the initial surges, make particularly attractive data from imaging radars with respect to optical sensors. Synthetic aperture radars are mostly insensitive to cloud cover and also to sun illumination. The third type of application is the one which involves the most advanced processing techniques. Remote sensing can play here a key role in defining the spatial and temporal changes of factors, which, in return, control flood generation and risk. For instance, data and algorithms are very promising in defining two seldom defined features of flooding: one is the spatiotemporal evolution of the flood. This often relies mainly on eyewitness accounts, especially where no stream gauge measurements are available. Accounts from eyewitnesses are occasionally cited in flash flood studies, and they have seldom been, to our knowledge, systematically collected and analysed. Multi-temporal and multisensor data allow a temporal and spatial reconstruction of flood inundation, from the beginning until the end when all the inundated areas return dry. The second is related to the sediment transfer processes, as, for example, erosion and deposition on slopes and in river beds, hyperconcentrated mud, debris flows, etc. Erosion and sedimentary features mainly focus on the post-flood investigation and constitute an indicator of the local runoff generation. Those features associated with flooding have been documented in both modern and past cases [33]. However, only a few studies have demonstrated the relationships between these features and the corresponding hydraulic conditions that produced them, making it difficult to evaluate, for example, the magnitude of palaeo-flooding. Reconstructions of palaeo-flood hydrology are needed to improve prevention strategies to mitigate future flooding events. The accuracy of such reconstructions is enhanced by the detailed studies of modern analogues [109].

In this respect, multi-temporal and multisource RS can play, in our view, a central role in filling the gap between these two different trends: the hydrological and statistical approach, mainly based on global river and flood inundation models, and the "geomorphological understanding" of floods based on ground effect evidence of actual and historical events.

Several reviews on the use of remote sensing in flood monitoring have been [46, 94] and are being published at the time of this writing, e.g. [88, 101], so the reader is referred to such works and the references therein, to get a better idea of the many aspects of flood monitoring through RS.

3 Optical Data

Optical remote sensing data are broadly defined as data provided by sensors in the spectral range going from short-wave infrared, through the visible range, up to the ultraviolet [76]. As mentioned above, optical data constitute a very useful source of data about flood extension and evolution. Clean water surfaces generally absorb most of the incident e.m. energy, except in the specular direction. Therefore, they are easily recognizable as "dark" areas in optical images acquired in geometries far from sunglint. The separation becomes less immediate as one considers water less clean, i.e. with increasing quantities of suspended sediments, or in case of increased reflections from sunlight due to acquisition geometries with shallow incidence angles or to the presence of wind roughening the water surface, thus increasing the chance of local specular reflections. In some such cases, recurring to several spectral bands can help. The main obstacle to the use of optical data for flood monitoring comes from their sensitivity to the presence of clouds, which usually are present during flood events. This factor alone hinders, to date, a widespread adoption of optical data within operational flood monitoring contexts.

3.1 Multispectral Data

When cloud cover or sun illumination does not preclude data acquisition in the optical domain, floodwaters can be detected with good precision by exploiting several typical characteristics of inland water surfaces with respect to dry areas.

One is the reduced reflectance of clean and calm water with respect to land areas. This behaviour is common to virtually all the optical spectral range, as long as the acquisition is far from the specular direction. An additional way to distinguish the presence of surface water is given by the availability of reflectance information in the infrared thermal spectral bands. This is also generally low for water surfaces, with respect to land areas.

Both these methods rely on assumptions which are broadly fulfilled when monitoring flood events occurring over non-urban land areas, especially when using low- [7] or medium-resolution data [48]. However, it is not uncommon to find local violations, especially with high-resolution data [30]. For instance, artificial surfaces may have very low reflectances and thus be mistaken for flooded areas. This is more likely to occur on highly complex environments like urban areas, where pavements and tarmacs may have a wide range of reflectance behaviours during a flood event. This may also include flooded surfaces exhibiting artificially high reflectance, which could be due, for instance, to a shallow water layer over a bright pavement or to specular reflections from surrounding buildings. Such cases could be successfully solved by sensors having infrared detection bands.

In fact, urban areas are admittedly among the most complicated types of land cover for many remote sensing applications, in virtually all regions of the e.m. spectrum, with flood monitoring making no exception (see also the radar applications, later on).

3.2 Vegetation and Water Indices

Whenever multispectral information is available, a compact way of recording effectively the presence or absence of water (or other land cover types) on the ground is through the use of indices, intended broadly as quantitative comparisons (usually ratios or differences) between the response of each ground pixels in different bands of the e.m. spectrum.

Several such indices have been devised, such as the Normalized Difference Water Index (NDWI), defined as a normalized ratio of the green and infrared reflectances [69], or other variously defined water indices (WI) [45, 91]. Their purpose is to enhance the sensitivity to the presence of inland water surfaces by taking into account multiple spectral bands. Many of these indices have been tested in flood contexts [16, 70].

Comparing the red and infrared bands is considered the most effective way to detect vegetation, combined in the so-called Normalized Difference Vegetation Index, or NDVI [93], which is one of the best known and widely used spectral indices. NDVI decreases on non-vegetated terrain, which has more uniform responses across red and infrared bands. On the other hand, water emission in the infrared is generally lower than in the red part of the spectrum, so water has an inverse NDVI behaviour with respect to both vegetated and unvegetated land. This makes NDVI a suitable tool to detect rapidly water surfaces.

4 Synthetic Aperture Radar (SAR) Data

Synthetic aperture radar data are a precious resource in flood monitoring. As explained above, the most common situation during a flood event is the widespread presence of clouds over the region, often for several days, which precludes acquisition with optical sensors. In this case, microwaves can be exploited, thanks to their longer wavelength which makes them propagate through clouds. This, together with the active nature of radar sensors, i.e. the fact that the sensor emits the radiation which is then collected after scattering from the surface, makes radar the only viable means to obtain flood maps with reliable time schedules.

Radar remote sensing has a long history, dating back to the years immediately after World War II and reaching maturity in the 1970s [105]. The possibility of controlling microwave emissions through antennae, in both amplitude and phase, led to the development of synthetic aperture techniques [18], so that synthetic aperture radar (SAR) is the most used technology for imaging radars nowadays [20].

The possibility of coherently controlling microwave emission and detection soon led to conceive the possibility of combining two coherent SAR acquisitions in an interferogram [6], in which the chaotic phase changes from pixel to pixel due to coherent superposition of e.m. returns from all the single scattering objects in a resolution element cancel, evidencing the contribution of the optical path difference to the interferometric phase [40].

Application of SAR and interferometric SAR (InSAR) to flood monitoring dates back to the 1990s [34, 121], with some systematization in subsequent years [25, 74]. More recently, SAR sensors have achieved unprecedented resolutions and repetitivity of acquisition, so that their application to flood monitoring is receiving renewed attention.

Several summaries and reviews are being published in the last few years regarding state-of-the-art applications of SAR to flood monitoring, e.g. [50, 56, 101]. Very often review works include SAR together with other sensor or technologies [46, 73], mainly because of the objective difficulties in detecting water on some types of terrain or in certain conditions, as better described in the following subsections.

In the next sections, we analyse separately the contribution of SAR backscatter and InSAR coherence to the purpose of flood monitoring.

4.1 SAR Backscatter Analysis

Radar applications to flood monitoring are based on the physical characteristics of the water surfaces in contrast to land areas, relying on the common principle that smooth water acts basically as a mirror for millimetric e.m. waves, backscattering most impinging radiation in the specular direction. Since this direction is usually away from the receiving radar antenna, water-covered areas appear darker than land areas. This contrast allows, in principle, to distinguish flooded from non-flooded areas. This condition can be generally assumed as verified in many cases. Reasons for non-compliance include wind, which may roughen the water surface, or the presence of vegetation, as well as surrounding walls or other man-made structures, causing reflections. All these effects may cause backscatter returns from flooded areas to increase, rather than decrease, with respect to dry areas. If such cases can be excluded, flood monitoring from SAR reduces to rather simple image processing steps, essentially consisting in thresholding of backscatter levels.

Literature about SAR backscatter applications to flood monitoring is becoming more and more consistent as attention to remote sensing increases. In [59], a comprehensive analysis of different backscatter levels in flooded and non-flooded conditions is presented for several land cover classes. In [101] a review of recent achievements in SAR flood monitoring is offered, starting from low-resolution, global initiatives up to the most recent experiments with high-resolution SAR systems. In [11] a sample use of airborne SAR data is shown on urban areas (Severn River, UK).

Several conclusions can be learned by studying such literature. Many are discussed and summarized in some of the chapters of the present book. Some common points which can be noticed are the following:

- Due to the general shape of the e.m. scattering pattern, for monostatic radars,[1] high backscattering from smooth surfaces is less and less likely as incidence angles[2] increase, so that also contrast between land and (calm) water surfaces increases with incidence angles. In practice, then, large incidence angles are better than smaller ones for discriminating water from other types of land cover [58, 77].
- HH polarization is preferred over VV or cross-polarized channels (VH or HV),[3] again because of contrast considerations: in fact, VV polarization results to be less sensitive to the presence of water than HH polarization, while cross-polarized channels exhibit generally more noise than co-polarized ones and are thus less efficient [57, 75, 114], although combination of co- and cross-polarized channels can improve discrimination capabilities of threshold-based methods [41].
- Discrimination of flood over bare terrain surfaces is hindered by conditions of increased return from water, e.g. in the presence of waves [81] or ice [82], which increase the roughness of the water surface.

[1] In a monostatic, active remote sensor, the receiving antenna is placed close (possibly coincides with) the transmitting one. Otherwise, the sensor is said to be bistatic.

[2] The incidence angle is defined as the angle between the direction of observation of the sensor (also known as line of sight or LOS) and the vertical direction.

[3] As coherent devices, the polarization of waves emitted and received by radar sensor antennae can be controlled. Most currently available SAR instruments work with linear polarizations, conventionally referred to the horizontal (H) or vertical (V) direction. Transmitted and received polarization can be controlled independently, so that co-polarized (H-transmitted, H-received, or HH, and V-transmitted, V-received, or VV), as well as cross-polarized, channels (VH or HV) can be investigated [76, p. 541].

- Cropland and urban areas are the most "difficult" land cover types, due to the widespread presence of vertical structures – vegetation for cropland, buildings and other man-made structures for urban areas – causing spurious reflections, double bounce, layover and other effects increasing returns from flooded areas with respect to dry ones [36, 62, 83].
- Forest canopies and wetlands are other types of land cover with good chances of discrimination [111], especially when long wavelengths are used (e.g. L or P band [42]).[4]
- An important factor limiting the use of SAR data at short wavelengths (X band) is the possible influence of clouds on the backscatter images [60, 61].
- Performance with respect to wavelength is difficult to predict [101]: sensitivity to surface roughness increases with frequency, so contrast with dry land may decrease X-band SAR intensity images compared to lower frequencies, in case of wind-roughened flooded areas. On the other hand, higher penetration may increase the chance of double-bounce effects [42] leading to missed alarms for flood detection. Other factors involve the increased sensitivity of higher frequencies to atmospheric phenomena like rain or snow [85], as well as the reduced effects of temporal interferometric decorrelation for L band compared to C or X bands, which leads to increased contrast between land and water in InSAR coherence images and thus generally better detection [78].

Thanks to their aforementioned desirable characteristics, especially their reliable availability, if properly planned, regardless of atmospheric or illumination conditions, SAR-derived flood maps look promising for integration with hydraulic models, which require regularly acquired external data for calibration or general performance assessment [66, 67, 113] (see also Sect. 8).

SAR data, like all coherent imaging devices, are affected by speckle, which is a high-frequency modulation of the detected intensity pattern, due to the coherent nature of radar sensors. Speckle tends to be considered as noise, since this high-frequency variation pattern is seen as superposed (multiplied) to an underlying lower-frequency variation of the radar backscatter cross-section, linked to different kinds of terrain. This is definitely the case in flood mapping applications, where the typical purpose is to have a clear and physically consistent indication of the areas affected by water. Speckle is considered in this case a nuisance and has to be filtered out of the data. This can be done as a preprocessing step or, in some cases, through post-processing of the mapping results.

SAR speckle filtering has become a wide and complex research field in itself (see, e.g. a comprehensive review in [3]). One of the main issues is the multiplicative nature of the speckle noise, which makes less effective many "traditional" filters, used in image processing, which are designed for use with imagery affected by

[4]Radar frequency ranges are divided into bands, whose names and limits derive mainly from historical military conventions [105]. The most used by current satellite radar sensors are L (1 to 2 GHz), C (4 to 8 GHz) and X (8 to 12.5 GHz) bands, with P (<300 MHz) and, on the other extreme, K bands (>12 GHz) mostly used in airborne sensors.

additive noise types. Many efforts have been devoted to devise SAR speckle adaptive filters, which make use of the local statistics of the SAR signal itself, to avoid averaging out localized features, thus retaining image resolution over highly variable regions while allowing homogeneous areas to be averaged more heavily to improve SNR. Among such filters, the most used and known are those by Lee [51], Frost [32] and Kuan [49]. Many more, including many recent improvements on classical filters and others based on other assumptions and principles, have become recently available. Their description, together with references, can be found in review works, such as [3, 72].

We just mention a particular kind of approach to speckle filtering, the so-called nonlocal approach [24], which has been applied with success to multidimensional SAR/InSAR data [23] and is gaining more and more attention in the SAR community, spawning a considerable amount of research [27, 55, 115].

4.2 Flood Detection Through InSAR

SAR interferometry is a technique developed with the aim of inferring 3-D information from pairs of SAR images (see introductory works, e.g. [6, 92]). InSAR data exhibit a very high sensitivity to ground features, allowing measurements of topographic height with accuracy of a few metres and displacements (through removal of topographic information, e.g. via use of external DEMs) with millimetric accuracies. Information about the topographic height or its displacements is contained in the phase of an interferogram, which can be approximated as $\Delta\phi = \phi_2 - \phi_1 \propto \frac{4\pi}{\lambda}\Delta r$, where ϕ_1 and ϕ_2 are the phases of the two complex SAR images composing the pair, λ is the radar wavelength, and Δr is the radar path delay difference between a given pixel and the sensor position in the two images.

An important contribution of the InSAR technique to the study and monitoring of floods is given by this capability of retrieving accurate topographic information over large areas. Besides local efforts, widespread use of InSAR-derived topographic information started with the availability of digital elevation data over most of land areas obtained from data collected during the NASA Shuttle Radar Topography Mission (SRTM), which flew in 2000 aboard the Space Shuttle Endeavour [29]. Distributed originally at a nominal posting of 3 arcsec (about 90 m at medium latitudes) [87], with several post-processing phases to ensure hydrologic consistency, and recently at 1 arcsec [103], SRTM data remain to date the most used topographic product for medium-resolution applications. A second-generation global DEM has been acquired by the TanDEM-X mission, flown by the German space agency (DLR). The derived global, 12.5 m posting DEM product is being completed and checked for distribution [122].

The level of reliability of the information contained in the phase is given, as in all interferometric applications, by the coherence of the two interfering radiation fields, I_1 and I_2. Interferometric coherence is defined as the complex number [15]:

$$\gamma = \frac{\langle I_1 I_2^* \rangle}{\sqrt{\langle I_1 I_1^* \rangle \langle I_2 I_2^* \rangle}}, \tag{1}$$

where the asterisk indicates complex conjugation. The average in the previous formula is intended as an ensemble average, operated over a homogeneous sample. In practice, it is often substituted by a summation over a finite image window W, i.e. the following definition of a sample coherence is used [110]:

$$\tilde{\gamma} = \frac{\sum_{p \in W} \left[I_1(p) I_2^*(p) \right]}{\sqrt{\sum_{p \in W} \left[I_1(p) I_1^*(p) \right] \sum_{p \in W} \left[I_2(p) I_2^*(p) \right]}}. \tag{2}$$

The $\tilde{\gamma}$ estimator is sensitive to the pixel-wise difference of the complex backscatter patterns of the two images I_1 and I_2, so that any subtle variation of this pattern between the two images composing the pair causes its value to decrease. Incidentally, this imposes that the two images be coregistered to within a fraction of a pixel in order for the estimator to bring reliable information [40]. As long as this precise coregistration is ensured, any variation in the complex patterns of the images will be recorded by the interferometric coherence.

In the particular case of a flooded surface, it is easy to realize that coherence will show invariably low values over surfaces which have undergone a flood wave anytime between the two acquisitions. As long as the passage of water causes long-standing variations in the micro-scale surface patterns, coherence will be low even in case neither the first nor the second image is acquired in the presence of standing water surfaces, but the area has experienced the passage of water at some time during the interval between the acquisitions. Provided this interval is not too long, the low-coherence areas will be recognizable with respect to other areas not affected by the flood, which will likely show higher coherence values. In practice then, if the above conditions are met, the coherence information acts as a sort of persistent change detector, registering the maximum extent of inundated areas between the two acquisitions of the interferogram [74].

Cases where the above general conditions are not fulfilled are of course those where other factors cause low-coherence values on dry surfaces as well as flooded ones, e.g. in the presence of vegetation or other land cover types exhibiting fast decorrelation times. In such cases, discrimination of flooded areas can become cumbersome even with short-time interferometric pairs [83, 89].

4.3 Electromagnetic Modelling

Most of the considerations sketched above about SAR intensity and InSAR coherence regard situations in which SAR backscatter from the Earth surface can be interpreted in a somewhat heuristic fashion, i.e. in terms of higher or lower values of average backscatter intensity or InSAR coherence, over relatively homogeneous areas. This is true, e.g. for bare soils or similar kinds of land cover. Even though accurate knowledge of expected returns greatly simplifies data interpretation, usually visual inspection of suspect areas, although time-consuming, allows to infer the presence or absence of water. However, other types of surfaces which can be affected by flood, especially those having more economic, artistic or historical value, may show much more complicated backscattering and coherence patterns, both in absence and in presence of water. Consider, e.g. the case of urban areas, where the presence of buildings and infrastructures may cause geometric patterns involving layover, shadowing or extreme foreshortening of façades or other objects. In such environments, discrimination of floodwaters may become extremely difficult [62, 65]. Some aid may come from detailed electromagnetic models of the response of such surfaces to impinging electromagnetic waves. Electromagnetic models have been developed during the last decades in a research field of its own [112]. For the specific field of flood monitoring, modelling and simulation have had an undiscussed role in helping understand some issues with enhanced backscatter below vegetation canopies [90, 118, 119] or in urban environments [5, 31].

As better detailed in one of the chapters of this book, in some cases the radar backscatter from inundated ground can be so different from what one expects in simple situations – in terms of absolute levels, spatial patterns and/or temporal variations – that e.m. modelling becomes necessary to understand controversial issues and thus try to reliably recognize flooded surfaces.

5 Time Series Interpretation

One of the most innovative aspects of latest generation sensors is their much increased capacity of acquiring images of the earth surface at regular, short intervals, with respect to previous missions. This constitutes a formidable step forward in the acquisition of data for the study of surface processes [39]. For the specific case of flood monitoring, the availability of time series of remotely sensed images constitutes an advantage from at least two points of view.

First, availability of frequent, detailed terrain maps for areas known to be periodically interested by inundation phenomena allows to infer precious information about the nature and genesis of the recurrent floods. This is true, for instance, for the comprehension of phenomena occurring on wetlands [96]. Also, long time series of detailed remotely sensed maps are very useful in the performance assessment and calibration of hydraulic models [99]. An important field in this respect is that covered by passive microwave instruments, which, in spite of their generally low spatial resolution, have very high acquisition frequencies, up to hours, so that very detailed time series of river discharge, and thus anomalies thereof preluding to floods, can be acquired [22, 108].

Second, for the analysis of specific events, the availability of large quantities of data previous (or posterior) to the flood helps in shading light on the land cover characteristics of "difficult" areas such as crops or urban areas, so that precision in mapping can be increased [21, 84]. Also, following the post-flood water decline may help understanding terrain evapotranspiration and absorption characteristics.

6 Methods to Produce Accurate Flood Maps

Together with remotely sensed data, effective methods to exploit their information are needed. These typically rely on methods for identifying different categories associated to complex land surfaces. In this respect, classification is one of the major approaches to infer land cover information from remotely sensed data [68].

Actually, the problem of flood detection from remotely sensed images can be considered as a particular instance of the more general topic of the detection of land cover changes and transitions [104]. As basic principle, change detection involves the analysis of at least two remote sensing images acquired over the same geographical area at different times: for example, assuming that a flood occurred in a given area at time t^*, two images acquired, respectively, at time $t_0 < t^*$ (before-event image) and $t_1 \geq t^*$ (co- or post-event image) are needed.

Traditionally, different pattern recognition methods and machine learning algorithms have been used to automatically classify multi-temporal remotely sensed images and to extract change information [17, 53]. In general, they can be divided into:

- Supervised or semi-supervised techniques that allow to detect land cover transitions but require an adequate amount of labelled samples to train the classifier;
- Unsupervised techniques that aim at discriminating between two opposite classes (representing changed and unchanged areas) without any a priori knowledge about the scene.

In the former case, data corresponding to image pixels for which the land cover class is known in advance are mandatory. Operationally, these samples may come from different sources, but they are often difficult and onerous to be collected, also considering that multi-temporal information is needed. Moreover, flooded area detection is usually needed in rapid mapping procedures to be performed shortly after the event. For these reasons, in the mapping of flooded areas, only few examples of application of supervised classifiers are reported in literature [37, 44, 65].

More frequently, the flood detection problem has been tackled as a clustering or a segmentation problem. Clustering is an unsupervised pattern recognition technique that partitions a set of n unlabelled patterns into K groups, on the basis of some similarity/dissimilarity measures, where the value of K may or may not be known a priori [13]. These patterns can be the information directly extracted by the images or indirect change indicators. Different unsupervised procedures have been reported in the literature [12, 36, 84, 107]. Some examples are illustrated in subsequent chapters of this book.

Flood maps derived from remotely sensed images are always, although to different extents, affected by errors, no matter what type of images and which pattern recognition algorithm have been used. To improve the accuracy of these maps, different data fusion approaches have been proposed, based on the synergetic use of different types of images and ancillary information. In the recent literature, several examples of data fusion methods applied to flood monitoring can be found. Surely, different factors have contributed to the strong interest in this topic, such as the recent programmes of satellite data acquisition, which have received a significant boost with the launch of new sensors, the new policies for data distribution of archive as well as of newly acquired satellite data (e.g., the ESA Sentinels). Together with the increased availability of multi-temporal, multisensor, multi-resolution and multiplatform imagery data, suitable algorithms to manage these information have to be considered. For example, the 2009–2010 Data Fusion Contest organized by the Data Fusion Technical Committee of the IEEE Geoscience and Remote Sensing Society was focused on the detection of flooded areas using multi-temporal and multimodal images. Both high spatial resolution optical and SAR data were provided, together with a digital elevation model. The goal was to investigate the improvement derived from decision fusion. As expected, the best map is the one obtained by using all the available data, the accuracy of which outperforms other ones, obtained by considering only subsets of the data [52].

Many other applicative examples are available. In [67], two different statistical methods have been used to integrate SAR data with high-precision topography and a river flow model. In [102], two experimental studies about the synergetic exploitation of remote sensing data and hydrometeorological models have been presented. More recently, flood maps have been produced combining multi-temporal SAR data and a hydrodynamic model by using a classification algorithm based on fuzzy logic [86], and a procedure has been developed in order to estimate the flooded area combining SAR imagery with ancillary information extracted from digital elevation topography [14]. In this book, an application of a data fusion model, based on Bayesian networks, is presented and discussed.

The last frontier of flood detection algorithms is their transformation into operational services that are expected to:

- Usefully exploit the large amount of available imagery data and their high temporal and spatial resolution, by fully exploiting the paradigms of the big data analysis [54];
- Enhance the detection by merging remotely sensed imagery data with other information, such as the one extracted by networks of ground or mobile sensors, connected through the Internet or dedicated links;
- Operate in a near-real-time mode, also allowing to detect flood events lasting a few days or hours and/or involving small areas (flash floods).

Nowadays, despite the increasing number of fully automated image processing algorithms, only a few operational flood services, based on remotely sensed image exploitation, exist [59]. For instance, the Fast Access to Imagery for Rapid Exploitation (FAIRE) service, hosted on the European Space Agency's (ESA) Grid

Processing on Demand (G-POD) system, provides automatic SAR preprocessing and change detection capabilities which can be triggered on demand by a user via a web interface [19]. The application is currently being extended with mapping capabilities which are based on a comparison of SAR data acquired during flood situations with corresponding archive/reference data acquired at normal water levels. More recently, a prototype automated processing algorithm for medium-resolution surface water mapping based on systematic and global scale ENVISAT ASAR acquisitions has been presented [120]. The proposed technique, which is embedded in an online service, classifies open water areas according to the probability of each image element to be covered by water according to 2-D histogram training datasets built from archive data in the incidence backscatter domain. A further example is the Fully Automatic Aqua Processing Service (FAAPS) which aims at developing a future operational service delivering flood extent maps generated from ESA satellite data. The methodology is based on a priori knowledge about an area taken from a multi-temporal time series and digital elevation information [98].

7 Validation

Validation of flood maps remains one of the most critical aspects of the entire monitoring efforts. To objectively validate a map or a mapping approach, it is necessary to have independently derived data about flood extension and perform some statistical comparisons between the proposed map and such ground-truth data. However, due to all the problems cited in the preceding parts of this chapter, having reliable ground-truth flood data over meaningful spatial extents is at present considered difficult, and only a few cases are known in which this has been possible. The most typical situation in which this comparison is possible, although invariably on limited areal extents, is in case of availability of pictures or videos taken in situ by eyewitnesses, in the immediate period after a flood event [83]. This is in fact the traditional way in which independent ground truth is meant (see also considerations in Sect. 2). It is easy to realize that such kind of data are fairly rare. Yet, they are among the most useful data to assess objectively remote sensing-based approaches.

As a consequence of the difficulty in finding valuable ground truth, indirect validation approaches are often adopted, such as comparison of data derived from independent remotely sensed sources (e.g. radar vs. optical) or from independent approaches (remote sensing vs. hydrological modelling). Such attempts must necessarily consider the different nature of the data which are being compared, to properly take into account possible biases, different assumptions and other distortions. In particular, when comparing two different mapping methodologies, uncertainties connected to both sides of the experiment should be kept in mind, rather than considering one approach as "ground truth" and assess the other one with respect to the former. This is particularly difficult in cases when the two approaches come from knowledge fields traditionally distant from each other. Nevertheless, when properly considered, often such comparisons may give a good idea of how realistic is a mapping approach, provided this comparison is considered for its actual value, i.e. basically qualitative.

A recent step forward in the direction of objective validation is the realization of the possibility of applying rigorous statistical techniques to attach probabilities to remote sensing-derived flood maps [21, 35, 97]. This allows to produce maps reporting, e.g. the probability of each pixel or image object being actually flooded at a particular date, based on knowledge coming from remotely sensed imagery and other data. Such probabilistic maps can be used in several ways, e.g. by converting them to binary maps through hard thresholding or by retaining the uncertainty information in order to assess the capability of a given mapping methodology to efficiently detect flooded areas. Rigorous probabilistic methods are indeed used currently in model assessments and calibration through remotely sensed data [10, 47] (see Sect. 8).

In spite of these recent advances, however, in many cases, none of the above-mentioned methods are applicable, since simply no valid independent ground truth is available. This happens especially in operational real-time or near-real-time contexts or when few data are available for a given event (e.g. a single SAR or optical image). In such cases, it is not uncommon to adopt heuristic, qualitative methods of validation for remotely sensed flood maps, mainly based on simple rules and common sense.

8 Hydrological Modelling

A pair of recent commentaries by P. Bates set the state of the art for the use of remotely sensed data in hydrological modelling [8, 9], monitoring at a distance of 8 years the advancements in the discipline. Flood inundation research has in the last years shifted from being a "data-poor" to a "data-rich" scientific field, mainly thanks to remote sensing data [9]. There is now a consensus between remote sensing professionals and hydrologists on the importance of integrating remote sensing-derived data and hydrological modelling [64]. Several studies have explored the potential of this integration, and significant progress has been achieved, especially on how high-resolution SAR images can support or also advance hydrological modelling (e.g. [38]). Studies on integrating remotely sensed data with flood modelling are dated back to the late 1990s, and most of them have demonstrated a great effectiveness in combining satellite-derived flood maps with digital elevation models and distributed process models. In fact, integrating inundation maps with topography can reveal complexities in water surface dynamics and in local flow connectivity which are otherwise difficult to model numerically or observe with remote sensing data alone. Despite this great effort, uncertainty in observations, as well as model parameters and predictions, is still a major challenge in fluvial hydrology research [26].

Many research papers have explored the integration of remote sensing with hydrological flood modelling covering several aspects; most of them focus on four major challenges (see e.g. [99]):

- The use of remote sensing data to retrieve and evaluate flood hydrology information, such as flood area extent, water stage, discharge, etc.;
- The use of remote sensing imagery to retrieve hydro-geomorphic parameters (e.g. roughness parameters, river cross-section geometries, detailed floodplain topography), which can be used to build model structures and to enhance process representation in the model;
- The use of remote sensing observations to calibrate and validate hydrodynamic models;
- The use of remote sensing information in terms of data assimilation with models.

One of the most common applications deals with the retrieval and mapping of inundation area extent and other flood hydrology information, since hydrological data collection is still considered a difficult task to date [106]. Inundation maps have been traditionally drawn from aerial photography and, despite the high costs, this is still one of the most reliable methods for accurate mapping. However, as shown in the preceding sections, the recent availability of high-resolution satellite imagery makes remote sensing-derived data very keen to flood monitoring and mapping. In particular, the strengths of spaceborne radar imaging and its peculiar characteristics, such as penetration of clouds, insensitivity to weather and possibility of data acquisition during day and night, make them very useful for the monitoring and mapping activities [89]. To map flood extent with optical images, the common mapping techniques consist of digitizing the boundaries at the contrasting land-water interface. Another approach involves the use of image processing for the extraction of a binary map consisting of dry and wet (flooded) pixels using different types of algorithms and bands. However, both approaches present drawbacks related to the possible presence of cloud cover and to the inability to map inundation beneath the vegetation canopies. For these reasons, active microwave imagery from SAR has gained momentum as the most reliable source of information to map flood extents. Besides inundation mapping, aerial photo, optical and SAR images have been used to indirectly derive water stage, combining the elevation of the land-water interface with digital elevation models or topographic maps. In particular, thanks to the availability of high-resolution DEMs (e.g. LiDAR data[5]), which are becoming more readily available, and the flood boundaries extracted from high-resolution SAR or optical images, it is possible to derive water stage and flow depth across the floodplain for a given event.

Hydrological and geomorphologic data retrieval from remotely sensed images has also advanced the integration with hydraulic modelling. For example, in [106],

[5]Light detection and ranging (LiDAR) sensors collect large clouds of precisely geocoded points, which can be then fused together into high-resolution matrices or sparse representations of the Earth surface, providing the highest-resolution DEMs available to date [116].

a high correlation between "width" (inundation area) and discharge in braided rivers was found. It is thus suggested that the relationship can be applied to estimate hydrologic variables from space. However, a central role is still played by high-resolution LiDAR data, which can result in highly detailed topography of the entire floodplain, on which hydraulic models are built and run. LiDAR-derived vegetation height has been also used to derive roughness estimates, to be included as floodplain friction parameterization in two-dimensional flood models. Roughness is a parameter that strongly influences model results and is usually poorly constrained.

Another important hydrological application of remote sensing-derived data is the calibration and validation of hydrodynamic models. There are many ways to evaluate the performance of flood inundation models in predicting flood extent or stage. However, the choice of the method depends on the details of the model, the type and nature of the observed data and the calibration or validation exercise. Several of the above-mentioned hydro-geomorphic types of information have been used to calibrate and validate model parameters and to test model performance. The most common method to evaluate model results is still the visual comparison by an expert flood modeller, who checks for consistency between model results and remotely sensed data, evaluating the hydraulic feasibility. Several papers propose quantitative measures aimed at a more objective assessment of model performance, mainly in terms of inundated area, and most of them are based on the comparison using a pixel-by-pixel approach. In particular, binary maps of wet/dry pixels are compared in terms of number of pixels correctly predicted as wet or dry, considering underprediction and overprediction. Receiver operating characteristic (ROC) curves are also used to evaluate the performance of the binary classified images, considering variable discrimination thresholds. More sophisticated methods, aimed at taking into account prediction uncertainties, apply fuzzy logic approaches to test predicted inundation maps, considering, for example, a membership function that includes assumptions of the probability of pixels being flooded [79, 80]. Studies that refer to the use of remotely sensed water levels in the model calibration and validation phase are still limited (see, e.g. [99]). Mason et al. [63] and Schumann et al. [100] estimated the mean cross-sectional water levels used for model evaluation from the intersection of SAR flood extent boundaries with a LiDAR DEM. The reason for the scarcity of examples of this kind of study can be found in the lack of high-resolution topographic data. This still remains an issue, although the increased availability of high spatial and radiometric resolution SAR data suggests a potential increase of this kind of studies.

The last application to integrate remote sensing-derived data and hydrological modelling is related to data assimilation into hydraulic modelling. Data assimilation is aimed at updating numerical models using both current and past observations, to provide temporal continuity and dynamic coupling among the variables. This kind of application is still in its infancy [117]. Most of the related works focus on the assimilation of remote sensing-derived water levels in flood forecasting systems (e.g. [2]). We point out the reader to a comprehensive analysis on land surface data assimilation by Houser et al. [43]. They show that assimilation of remotely sensed data can be achieved using basically four different classes of methods: (a) state updating methods to adjust internal model states based on observations, (b) input updating methods that generate new estimates from input data that are run through the simulation model, (c) error prediction methods that evaluate past errors between observations and model predictions to estimate future errors and (d) parameter updating methods, aimed at updating the parameters during the simulations.

9 Conclusions and Book Synopsis

In this chapter a summary of the most important aspects of detailed flood monitoring through remote sensing has been attempted. We reviewed the basics of flood monitoring practices, focusing on the commonly accepted standards for definition, detection and updating of flood maps. We then listed the main sources of data which are commonly used in flood monitoring activities, spanning through optical and microwave instruments, the main sources of remotely sensed data used in this field. We also tried to give a general idea of the main data processing techniques used to extract flood extent and evolution information from earth observation data and to integrate remotely sensed data into hydraulic inundation models, as well as to use remote sensing-derived information with the general purpose of improving hydro-geomorphic scenarios and enrich our knowledge of flood processes.

The rest of this book consists of a series of chapters covering some individual aspects of these activities. In Chap. 2, remote sensing data analysis of channel dynamics and geomorphic effects of floods are reviewed. In Chap. 3, the specific case of DEM data to derive useful information about recurrent or future flood events, as well as constrain hydraulic models for the identification of flooded areas, is considered. Chapter 4 is devoted to the field of flood forecast and real-time monitoring, with a review of some existing software solutions and online operational systems to rapidly provide information and forecasts about flood events. In Chap. 5, an interesting and rather unique test case, consisting of the study of a dataset composed of many radar and optical acquisitions, is described. Chapter 6 starts a suite of three chapters dealing with the most useful type of data for detailed flood monitoring, i.e. synthetic aperture radar data. In this case, important processing details are reviewed for simple combinations of single images acquired during the event or pairs of pre- and co- or post-event images. Chapter 7 is a very comprehensive review of most known aspects of SAR image interpretation linked to flood monitoring and detection in several land cover settings. Finally, Chap. 8 presents an innovative framework, based on Bayesian networks, which promises to help in dealing with the increasing flow of information from remote, in situ and other types of sensors, in order to extract relevant and unambiguous information about the development, in time and space, of inundation phenomena.

References

1. Alfieri, L., Salamon, P., Bianchi, A., Neal, J., Bates, P.D., Feyen, L.: Advances in pan-European flood hazard mapping. Hydrol. Process. **28**(13), 4067–4077 (2014). doi:10.1002/hyp.9947
2. Andreadis, K.M., Clark, E.A., Lettenmaier, D.P., Alsdorf, D.E.: Prospects for river discharge and depth estimation through assimilation of swath-altimetry into a raster-based hydrodynamics model. Geophys. Res. Lett. **34**(10), L10,403 (2007). doi:10.1029/2007GL029721
3. Argenti, F., Lapini, A., Bianchi, T., Alparone, L.: A tutorial on speckle reduction in synthetic aperture radar images. IEEE Geosci. Remote Sens. Maga. **1**(3), 6–35 (2013). doi:10.1109/MGRS.2013.2277512
4. Baker, V.R.: Geomorphological understanding of floods. Geomorphology **10**(1–4), 139–156 (1994). doi:10.1016/0169–555X(94)90013–2
5. Balz, T.: SAR Simulation of Urban Areas: Techniques and Applications, pp. 215–231. Springer, Dordrecht (2010). doi:10.1007/978-90-481–3751-0_9
6. Bamler, R., Hartl, P.: Synthetic aperture radar interferometry. Inverse Prob. **14**(4), R1–R54 (1998). doi:10.1088/0266-5611/14/4/001
7. Barton, I.J., Bathols, J.M.: Monitoring floods with AVHRR. Remote Sens. Environ. **30**(1), 89–94 (1989). doi:10.1016/0034-4257(89)90050–3
8. Bates, P.D.: Remote sensing and flood inundation modelling. Hydrol. Process. **18**(13), 2593–2597 (2004). doi:10.1002/hyp.5649
9. Bates, P.D.: Integrating remote sensing data with flood inundation models: how far have we got? Hydrol. Process. **26**(16), 2515–2521 (2012). doi:10.1002/hyp.9374
10. Bates, P.D., Horritt, M.S., Aronica, G., Beven, K.: Bayesian updating of flood inundation likelihoods conditioned on flood extent data. Hydrol. Process. **18**(17), 3347–3370 (2004). doi:10.1002/hyp.1499
11. Bates, P.D., Wilson, M.D., Horritt, M.S., Mason, D.C., Holden, N., Currie, A.: Reach scale floodplain inundation dynamics observed using airborne synthetic aperture radar imagery: data analysis and modelling. J. Hydrol. **328**(1–2), 306–318 (2006). doi:10.1016/j.jhydrol.2005.12.028
12. Bazi, Y., Bruzzone, L., Melgani, F.: An unsupervised approach based on the generalized Gaussian model to automatic change detection in multitemporal SAR images. IEEE Trans. Geosci. Remote Sens. **43**(4), 874–886 (2005). doi:10.1109/TGRS.2004.842441
13. Bishop, C.M.: Neural networks for pattern recognition. Clarendon Press/Oxford University Press, Oxford/New York (1995)
14. Boni, G., Ferraris, L., Pulvirenti, L., Squicciarino, G., Pierdicca, N., Member, S., Candela, L., Pisani, A.R., Zoffoli, S., Onori, R., Proietti, C., Pagliara, P.: A prototype system for flood monitoring based on flood forecast combined with COSMO-SkyMed and sentinel-1 data. IEEE J. Sel. Top. Appl. Earth Obs. Remote Sens. 1–12 (2016). doi:10.1109/JSTARS.2016.2514402
15. Born, M., Wolf, E.: Principles of Optics: Electromagnetic Theory of Propagation, Interference and Diffraction of Light. Pergamon Press, Oxford (1980)
16. Boschetti, M., Nutini, F., Manfron, G., Brivio, P.A., Nelson, A.: Comparative analysis of normalised difference spectral indices derived from MODIS for detecting surface water in flooded rice cropping systems. PLoS One **9**(2), e88741 (2014). doi:10.1371/journal.pone.0088741
17. Bovolo, F., Bruzzone, L.: The time variable in data fusion: a change detection perspective. IEEE Geosci. Remote Sens. Maga. **3**(3), 8–26 (2015). doi:10.1109/MGRS.2015.2443494
18. Cafforio, C., Prati, C., Rocca, F.: SAR data focusing using seismic migration techniques. IEEE Trans. Aerosp. Electron. Syst. **27**(2), 194–207 (1991). doi:10.1109/7.78293
19. Cossu, R., Schoepfer, E., Bally, P., Fusco, L.: Near real-time SAR-based processing to support flood monitoring. J. Real-Time Image Proc. **4**(3), 205–218 (2009). doi:10.1007/s11554–009-0114-4

20. Curlander, J.C., McDonough, R.N.: Synthetic aperture radar: systems and signal processing. Wiley, New York (1991)
21. D'Addabbo, A., Refice, A., Pasquariello, G., Lovergine, F.P., Capolongo, D., Manfreda, S.: A Bayesian network for flood detection combining SAR imagery and ancillary data. IEEE Trans. Geosci. Remote Sens. **54**(6), 3612–3625 (2016). doi:10.1109/TGRS.2016.2520487
22. De Groeve, T.: Flood monitoring and mapping using passive microwave remote sensing in Namibia. Geomat. Nat. Haz. Risk **1**(1), 19–35 (2010). doi:10.1080/19475701003648085
23. Deledalle, C.A., Denis, L., Tupin, F., Reigber, A., Jager, M.: NL-SAR: a unified nonlocal framework for resolution-preserving (Pol)(In)SAR denoising. IEEE Trans. Geosci. Remote Sens. **53**(4), 2021–2038 (2015). doi:10.1109/TGRS.2014.2352555
24. Deledalle, C.A., Loïc, D., Poggi, G., Tupin, F., Verdoliva, L.: Exploiting patch similarity for SAR image processing: the nonlocal paradigm. IEEE Signal Process. Mag. **31**(4), 69–78 (2014). doi:10.1109/MSP.2014.2311305
25. Dellepiane, S.G., Bo, G., Monni, S., Buck, C.: SAR images and interferometric coherence for flood monitoring. In: IGARSS 2000. IEEE 2000 International Geoscience and Remote Sensing Symposium. Taking the Pulse of the Planet: The Role of Remote Sensing in Managing the Environment. Proceedings (Cat. No.00CH37120), vol. 6, pp. 2608–2610. IEEE (2000). doi:10.1109/IGARSS.2000.859656
26. Di Baldassarre, G., Uhlenbrook, S.: Is the current flood of data enough? A treatise on research needs for the improvement of flood modelling. Hydrol. Process. **26**(1), 153–158 (2012). doi:10.1002/hyp.8226
27. Di Martino, G., Di Simone, A., Iodice, A., Riccio, D.: Scattering-based nonlocal means SAR despeckling. IEEE Trans. Geosci. Remote Sens. **54**(6), 3574–3588 (2016). doi:10.1109/TGRS.2016.2520309
28. European Council: EU directive of the European parliament and of the European council on the estimation and management of flood risks (2007/60/EU) (2007). http://ec.europa.eu/environment/water/flood_risk/
29. Farr, T.G., Rosen, P.A., Caro, E., Crippen, R., Duren, R., Hensley, S., Kobrick, M., Paller, M., Rodriguez, E., Roth, L., Seal, D., Shaffer, S., Shimada, J., Umland, J., Werner, M., Oskin, M., Burbank, D., Alsdorf, D.E.: The shuttle radar topography mission. Rev. Geophys. **45**(2), RG2004 (2007). doi:10.1029/2005RG000183
30. Feng, Q., Liu, J., Gong, J.: Urban flood mapping based on unmanned aerial vehicle remote sensing and random forest classifier – a case of Yuyao, China. Water **7**(4), 1437–1455 (2015). doi:10.3390/w7041437
31. Franceschetti, G., Iodice, A., Riccio, D., Ruello, G.: SAR raw signal simulation for urban structures. IEEE Trans. Geosci. Remote Sens. **41**(9), 1986–1995 (2003). doi:10.1109/TGRS.2003.814626
32. Frost, V.S., Stiles, J.A., Shanmugan, K.S., Holtzman, J.C.: A model for radar images and its application to adaptive digital filtering of multiplicative noise. IEEE Trans. Pattern Anal. Mach. Intell. PAMI **4**(2), 157–166 (1982). doi:10.1109/TPAMI.1982.4767223
33. Gaume, E., Borga, M.: Post-flood field investigations in upland catchments after major flash floods: proposal of a methodology and illustrations. J. Flood Risk Manage. **1**(4), 175–189 (2008). doi:10.1111/j.1753-318X.2008.00023.x
34. Geudtner, D., Winter, R., Vachon, P.: Flood monitoring using ERS-1 SAR interferometry coherence maps. In: IGARSS '96. 1996 International Geoscience and Remote Sensing Symposium, vol. 2, pp. 966–968. IEEE (1996). doi:10.1109/IGARSS.1996.516536
35. Giustarini, L., Hostache, R., Kavetski, D., Chini, M., Corato, G., Schlaffer, S., Matgen, P.: Probabilistic flood mapping using synthetic aperture radar data. IEEE Trans. Geosci. Remote Sens. 1–12 (2016). doi:10.1109/TGRS.2016.2592951

36. Giustarini, L., Hostache, R., Matgen, P., Schumann, G.J.P., Bates, P.D., Mason, D.C.: A change detection approach to flood mapping in urban areas using TerraSAR-X. IEEE Trans. Geosci. Remote Sens. **51**(4), 2417–2430 (2013). doi:10.1109/Tgrs.2012.2210901
37. Gnecco, G., Morisi, R., Roth, G., Sanguineti, M., Taramasso, A.C.: Supervised and semi-supervised classifiers for the detection of flood-prone areas. Soft Comput. (2016). doi:10.1007/s00500-015-1983-z
38. Grimaldi, S., Li, Y., Pauwels, V.R.N., Walker, J.P.: Remote sensing-derived water extent and level to constrain hydraulic flood forecasting models: opportunities and challenges. Surv. Geophys. **37**(5), 977–1034 (2016). doi:10.1007/s10712-016-9378-y
39. Guyet, T., Nicolas, H.: Long term analysis of time series of satellite images. Pattern Recogn. Lett. **70**, 17–23 (2015). doi:10.1016/j.patrec.2015.11.005
40. Hanssen, R.F.: Radar Interferometry: Data Interpretation and Error Analysis. Kluwer Academic Publishers (2001). doi:10.1007/s13398-014-0173–7.2
41. Henry, J.B.B., Chastanet, P., Fellah, K., Desnos, Y.L.L.: Envisat multi-polarized ASAR data for flood mapping. Int. J. Remote Sens. **27**(10), 1921–1929 (2006). doi:10.1080/01431160500486724
42. Hess, L.L., Melack, J.M., Simonett, D.S.: Radar detection of flooding beneath the forest canopy: a review. Int. J. Remote Sens. **11**(7), 1313–1325 (1990). doi:10.1080/01431169008955095
43. Houser, P.R., De Lannoy, G.J., Walker, J.P.: Land surface data assimilation. In: Data Assimilation, pp. 549–597. Springer, Berlin/Heidelberg (2010). doi:10.1007/978-3-540-74703-1_21
44. Ireland, G., Volpi, M., Petropoulos, G.: Examining the capability of supervised machine learning classifiers in extracting flooded areas from landsat TM imagery: a case study from a mediterranean flood. Remote Sens. **7**(3), 3372–3399 (2015). doi:10.3390/rs70303372
45. Ji, L., Zhang, L., Wylie, B.: Analysis of dynamic thresholds for the normalized difference water index. Photogramm. Eng. Remote Sens. **75**(11), 1307–1317 (2009). doi:10.14358/PERS.75.11.1307
46. Joyce, K.E., Belliss, S.E., Samsonov, S.V., McNeill, S.J., Glassey, P.J.: A review of the status of satellite remote sensing and image processing techniques for mapping natural hazards and disasters. Prog. Phys. Geogr. **33**(2), 183–207 (2009). doi:10.1177/0309133309339563
47. Jung, H.C., Jasinski, M., Kim, J.W., Shum, C.K., Bates, P.D., Neal, J., Lee, H., Alsdorf, D.E.: Calibration of two-dimensional floodplain modeling in the central Atchafalaya Basin Floodway System using SAR interferometry. Water Resour. Res. **48**(7), W07511, (2012). doi:10.1029/2012WR011951
48. Klemas, V.: Remote sensing of floods and flood-prone areas: an overview. J. Coastal Res. **314**, 1005–1013 (2015). doi:10.2112/JCOASTRES-D-14–00160.1
49. Kuan, D.T., Sawchuk, A.A., Strand, T.C., Chavel, P.: Adaptive noise smoothing filter for images with signal-dependent noise. IEEE Trans. Pattern Anal. Mach. Intell. PAMI **7**(2), 165–177 (1985). doi:10.1109/TPAMI.1985.4767641
50. Kussul, N., Shelestov, A., Skakun, S.: Flood monitoring from SAR data. In: NATO Science for Peace and Security Series C: Environmental Security, vol. 97, pp. 19–29 (2011). doi:10.1007/978-90-481-9618-0_3
51. Lee, J.S.: Speckle suppression and analysis for synthetic aperture radar images. Opt. Eng. **25**(5), 255636 (1986). doi:10.1117/12.7973877
52. Longbotham, N., Pacifici, F., Glenn, T., Zare, A., Volpi, M., Tuia, D., Christophe, E., Michel, J., Inglada, J., Chanussot, J., Du, Q.: Multi-modal change detection, application to the detection of flooded areas: Outcome of the 2009–2010 data fusion contest. IEEE J. Sel. Top. Appl. Earth Obs. Remote Sens. **5**(1), 331–342 (2012). doi:10.1109/JSTARS.2011.2179638
53. Lu, D., Weng, Q.: A survey of image classification methods and techniques for improving classification performance. Int. J. Remote Sens. **28**(5), 823–870 (2007). doi:10.1080/01431160600746456
54. Ma, P., Lin, H.: Robust detection of single and double persistent scatterers in urban built environments. IEEE Trans. Geosci. Remote Sens. **54**(4), 2124–2139 (2015). doi:10.1109/TGRS.2015.2496193

55. Ma, X., Shen, H., Zhao, X., Zhang, L.: SAR image despeckling by the use of variational methods With adaptive nonlocal functionals. IEEE Trans. Geosci. Remote Sens. **54**(6), 1–15 (2016). doi:10.1109/TGRS.2016.2517627
56. Manavalan, R.: SAR image analysis techniques for flood area mapping – literature survey. Earth Sci. Inf. (2016). doi:10.1007/s12145-016-0274-2
57. Manjusree, P., Prasanna Kumar, L., Bhatt, C.M., Rao, G.S., Bhanumurthy, V.: Optimization of threshold ranges for rapid flood inundation mapping by evaluating backscatter profiles of high incidence angle SAR images. Int. J. Disaster Risk Sci. **3**(2), 113–122 (2012). doi:10.1007/s13753-012-0011-5
58. Marti-Cardona, B., Lopez-Martinez, C., Dolz-Ripolles, J., Bladè-Castellet, E.: ASAR polarimetric, multi-incidence angle and multitemporal characterization of Doñana wetlands for flood extent monitoring. Remote Sens. Environ. **114**(11), 2802–2815 (2010). doi:10.1016/j.rse.2010.06.015
59. Martinis, S., Kersten, J., Twele, A.: A fully automated TerraSAR-X based flood service. ISPRS J. Photogramm. Remote Sens. **104**, 203–212 (2015). doi:10.1016/j.isprsjprs.2014.07.014
60. Marzano, F.S., Mori, S., Chini, M., Pulvirenti, L., Pierdicca, N., Montopoli, M., Weinman, J.A.: Potential of high-resolution detection and retrieval of precipitation fields from X-band spaceborne synthetic aperture radar over land. Hydrol. Earth Syst. Sci. **15**(3), 859–875 (2011). doi:10.5194/hess-15-859-2011
61. Marzano, F.S., Mori, S., Weinman, J.A.: Evidence of rainfall signatures on X-band synthetic aperture radar imagery over land. IEEE Trans. Geosci. Remote Sens. **48**(2), 950–964 (2010). doi:10.1109/TGRS.2009.2034843
62. Mason, D.C., Davenport, I.J., Neal, J.C., Schumann, G.J.P., Bates, P.D.: Near real-time flood detection in urban and rural areas using high-resolution synthetic aperture radar images. IEEE Trans. Geosci. Remote Sens. **50**(8), 3041–3052 (2012). doi:10.1109/TGRS.2011.2178030
63. Mason, D.C., Horritt, M.S., Dall'Amico, J.T., Scott, T.R., Bates, P.D.: Improving river flood extent delineation from synthetic aperture radar using airborne laser altimetry. IEEE Trans. Geosci. Remote Sens. **45**(12), 3932–3943 (2007). doi:10.1109/TGRS.2007.901032
64. Mason, D.C., Schumann, G.J.-P., Bates, P.D.: Data utilization in flood inundation modelling. In: Flood Risk Science and Management, pp. 209–233. Wiley-Blackwell, Oxford (2010). doi:10.1002/9781444324846.ch11
65. Mason, D.C., Speck, R., Devereux, B., Schumann, G.J.P., Neal, J.C., Bates, P.D.: Flood detection in urban areas using TerraSAR-X. IEEE Trans. Geosci. Remote Sens. **48**(2), 882–894 (2010). doi:10.1109/TGRS.2009.2029236
66. Matgen, P., Montanari, M., Hostache, R., Pfister, L., Hoffmann, L., Plaza, D., Pauwels, V.R.N., De Lannoy, G.J.M., De Keyser, R., Savenije, H.H.G.: Towards the sequential assimilation of SAR-derived water stages into hydraulic models using the Particle Filter: Proof of concept. Hydrol. Earth Syst. Sci. **14**(9), 1773–1785 (2010). doi:10.5194/hess-14-1773-2010
67. Matgen, P., Schumann, G.J.P., Henry, J.B., Hoffmann, L., Pfister, L.: Integration of SAR-derived river inundation areas, high-precision topographic data and a river flow model toward near real-time flood management. Int. J. Appl. Earth Obs. Geoinf. **9**(3), 247–263 (2007). doi:10.1016/j.jag.2006.03.003
68. Maulik, U., Chakraborty, D.: Remote sensing image classification: a survey of support-vector-machine-based advanced techniques. IEEE Geosci. Remote Sens. Maga. **5**(1), 33–52 (2017). doi:10.1109/MGRS.2016.2641240
69. McFeeters, S.K.: The use of the normalized difference water index (NDWI) in the delineation of open water features. Int. J. Remote Sens. **17**(7), 1425–1432 (1996). doi:10.1080/01431169608948714
70. Memon, A.A., Muhammad, S., Rahman, S., Haq, M.: Flood monitoring and damage assessment using water indices: A case study of Pakistan flood-2012. Egypt. J. Remote Sens. Space Sci. **18**(1), 99–106 (2015). doi:10.1016/j.ejrs.2015.03.003

71. de Moel, H., van Alphen, J., Aerts, J.C.J.H.: Flood maps in Europe – methods, availability and use. Nat. Hazards Earth Syst. Sci. **9**(2), 289–301 (2009). doi:10.5194/nhess-9-289-2009
72. Morandeira, N.S., Grimson, R., Kandus, P.: Assessment of SAR speckle filters in the context of object-based image analysis. Remote Sens. Lett. **7**(2), 150–159 (2016). doi:10.1080/2150704X.2015.1117153
73. Musa, Z.N., Popescu, I., Mynett, A.: A review of applications of satellite SAR, optical, altimetry and DEM data for surface water modelling, mapping and parameter estimation. Hydrol. Earth Syst. Sci. **19**(9), 3755–3769 (2015). doi:10.5194/hess-19-3755-2015
74. Nico, G., Pappalepore, M., Pasquariello, G., Refice, A., Samarelli, S.: Comparison of SAR amplitude vs. coherence flood detection methods – a GIS application. Int. J. Remote Sens. **21**(8), 1619–1631 (2000). doi:10.1080/014311600209931
75. Niculescu, S., Lardeux, C., Hanganu, J., Mercier, G., David, L.: Change detection in floodable areas of the Danube delta using radar images. Nat. Hazards **78**(3), 1899–1916 (2015). doi:10.1007/s11069-015-1809-4
76. Njoku, E.G. (ed.): Encyclopedia of Remote Sensing. Encyclopedia of Earth Sciences Series. Springer, New York (2014). doi:10.1007/978-0-387-36699-9
77. O'Grady, D., Leblanc, M., Bass, A.: The use of radar satellite data from multiple incidence angles improves surface water mapping. Remote Sens. Environ. **140**, 652–664 (2014). doi:10.1016/j.rse.2013.10.006
78. Oliver-Cabrera, T., Wdowinski, S.: InSAR-based mapping of tidal inundation extent and amplitude in Louisiana Coastal Wetlands. Remote Sens. **8**(5), 393 (2016). doi:10.3390/rs8050393
79. Pappenberger, F., Frodsham, K., Beven, K., Romanowicz, R., Matgen, P.: Fuzzy set approach to calibrating distributed flood inundation models using remote sensing observations. Hydrol. Earth Syst. Sci. **11**(2), 739–752 (2007). doi:10.5194/hess-11-739-2007
80. Pappenberger, F., Matgen, P., Beven, K.J., Henry, J.B., Pfister, L., Fraipont, P.: Influence of uncertain boundary conditions and model structure on flood inundation predictions. Adv. Water Resour. **29**(10), 1430–1449 (2006). doi:10.1016/j.advwatres.2005.11.012
81. Pierdicca, N., Pulvirenti, L., Chini, M.: Dealing with flood mapping using SAR data in the presence of wind or heavy precipitation. In: Notarnicola, C., Paloscia, S., Pierdicca, N. (eds.) Proceedings of SPIE, vol. 8891, p. 88910K (2013). doi:10.1117/12.2030105
82. Pultz, T., Leconte, R., St-Laurent, L., Peters, L.: Flood mapping with airborne SAR imagery: case of the 1987 Saint-John river flood. Can. Water Resour. J. **16**(2), 173–189 (1991). doi:10.4296/cwrj1602173
83. Pulvirenti, L., Chini, M., Pierdicca, N., Boni, G.: Use of SAR data for detecting floodwater in urban and agricultural areas: the role of the interferometric coherence. IEEE Trans. Geosci. Remote Sens. **54**(3), 1532–1544 (2016). doi:10.1109/TGRS.2015.2482001
84. Pulvirenti, L., Chini, M., Pierdicca, N., Guerriero, L., Ferrazzoli, P.: Flood monitoring using multi-temporal COSMO-skymed data: image segmentation and signature interpretation. Remote Sens. Environ. **115**(4), 990–1002 (2011). doi:10.1016/j.rse.2010.12.002
85. Pulvirenti, L., Marzano, F.S., Pierdicca, N., Mori, S., Chini, M.: Discrimination of water surfaces, heavy rainfall, and wet snow using COSMO-SkyMed observations of severe weather events. IEEE Trans. Geosci. Remote Sens. **52**(2), 858–869 (2014). doi:10.1109/TGRS.2013.2244606
86. Pulvirenti, L., Pierdicca, N., Boni, G., Fiorini, M., Rudari, R.: Flood damage assessment through multitemporal COSMO-SkyMed data and hydrodynamic models: the Albania 2010 case study. IEEE J. Sel. Top. Appl. Earth Obs. Remote Sens. **7**(7), 2848–2855 (2014). doi:10.1109/JSTARS.2014.2328012
87. Rabus, B., Eineder, M., Roth, A., Bamler, R.: The shuttle radar topography mission – a new class of digital elevation models acquired by spaceborne radar. ISPRS J. Photogramm. Remote Sens. **57**(4), 241–262 (2003). doi:10.1016/S0924-2716(02)00124-7
88. Rahman, M.S., Di, L.: The state of the art of spaceborne remote sensing in flood management. Nat. Hazards **85**(2), 1223–1248 (2017). doi:10.1007/s11069-016-2601-9

89. Refice, A., Capolongo, D., Pasquariello, G., D'Addabbo, A., Bovenga, F., Nutricato, R., Lovergine, F.P., Pietranera, L.: SAR and InSAR for flood monitoring: examples with COSMO-SkyMed data. IEEE J. Sel. Top. Appl. Earth Obs. Remote Sens. **7**(7), 2711–2722 (2014). doi:10.1109/JSTARS.2014.2305165
90. Richards, J.A., Woodgate, P.W., Skidmore, A.K.: An explanation of enhanced radar backscattering from flooded forests. Int. J. Remote Sens. **8**(7), 1093–1100 (1987). doi:10.1080/01431168708954756
91. Rogers, A.S., Kearney, M.S.: Reducing signature variability in unmixing coastal marsh Thematic Mapper scenes using spectral indices. Int. J. Remote Sens. **25**(12), 2317–2335 (2004). doi:10.1080/01431160310001618103
92. Rosen, P., Hensley, S., Joughin, I., Li, F., Madsen, S., Rodriguez, E., Goldstein, R.: Synthetic aperture radar interferometry. Proc. IEEE **88**(3), 333–382 (2000). doi:10.1109/5.838084
93. Rouse, W., Haas, R.H., Schell, J.A., Deering, D.W.: Monitoring vegetation systems in the Great Plains with ERTS. In: 3rd ERTS Symposium, pp. 309–317. NASA SP-351 I (1973)
94. Sanyal, J., Lu, X.X.: Application of remote sensing in flood management with special reference to monsoon Asia: a review. Nat. Hazards **33**(2), 283–301 (2004). doi:10.1023/B:NHAZ.0000037035.65105.95
95. Schanze, J., Zeman, E., Marsalek, J. (eds.): Flood Risk Management: Hazards, Vulnerability and Mitigation Measures. NATO Science Series, vol. 67. Springer, Dordrecht (2006). doi:10.1007/978-1-4020-4598-1
96. Schlaffer, S., Chini, M., Dettmering, D., Wagner, W.: Mapping wetlands in Zambia using seasonal backscatter signatures derived from ENVISAT ASAR time series. Remote Sens. **8**(5), 402 (2016). doi:10.3390/rs8050402
97. Schlaffer, S., Chini, M., Giustarini, L., Matgen, P.: Probabilistic mapping of flood-induced backscatter changes in SAR time series. Int. J. Appl. Earth Obs. Geoinf. **56**, 77–87 (2017). doi:10.1016/j.jag.2016.12.003
98. Schlaffer, S., Hollaus, M., Wagner, W., Matgen, P.: Flood delineation from synthetic aperture radar data with the help of a priori knowledge from historical acquisitions and digital elevation models in support of near-real-time flood mapping. In: Michel, U., Civco, D.L., Ehlers, M., Schulz, K., Nikolakopoulos, K.G., Habib, S., Messinger, D., Maltese, A. (eds.) Proceedings of SPIE, vol. 8538, p. 853813 (2012). doi:10.1117/12.974503
99. Schumann, G.J.P., Bates, P.D., Horritt, M.S., Matgen, P., Pappenberger, F.: Progress in integration of remote sensing-derived flood extent and stage data and hydraulic models. Rev. Geophys. **47**(3), RG4001 (2009). doi:10.1029/2008RG000274
100. Schumann, G.J.P., Matgen, P., Pappenberger, F., Black, A., Cutler, M., Hoffmann, L., Pfister, L.: The REFIX model: remote sensing based flood modelling. In: ISPRS Commission VII Mid-Term Symposium "Remote Sensing: From Pixels to Processes", p. 6. Enschede (2006)
101. Schumann, G.J.P., Moller, D.K.: Microwave remote sensing of flood inundation. Phys. Chem. Earth Parts A/B/C **83–84**, 84–95 (2015). http://dx.doi.org/10.1016/j.pce.2015.05.002
102. Serpico, S.B., Dellepiane, S.G., Boni, G., Moser, G., Angiati, E., Rudari, R.: Information extraction from remote sensing images for flood monitoring and damage evaluation. Proc. IEEE **100**(10), 2946–2970 (2012). doi:10.1109/JPROC.2012.2198030
103. Shuttle Radar Topography Mission (SRTM) 1 Arc-Second Global | The Long Term Archive. https://lta.cr.usgs.gov/SRTM1Arc
104. Singh, A.: Digital change detection techniques using remotely-sensed data. Int. J. Remote Sens. **10**(6), 989–1003 (1989). doi:10.1080/01431168908903939
105. Skolnik, M.I. (ed.): Radar Handbook, 2nd edn. McGraw-Hill, New York (1990)
106. Smith, L.C., Pavelsky, T.M.: Estimation of river discharge, propagation speed, and hydraulic geometry from space: Lena River, Siberia. Water Resour. Res. **44**(3), 1–11 (2008). doi:10.1029/2007WR006133
107. Su, L., Gong, M., Sun, B., Jiao, L.: Unsupervised change detection in SAR images based on locally fitting model and semi-EM algorithm. Int. J. Remote Sens. **35**(2), 621–650 (2014). doi:10.1080/01431161.2013.871596

108. Sulistioadi, Y.B., Tseng, K.H., Shum, C.K., Hidayat, H., Sumaryono, M., Suhardiman, A., Setiawan, F., Sunarso, S.: Satellite radar altimetry for monitoring small rivers and lakes in Indonesia. Hydrol. Earth Syst. Sci. **19**(1), 341–359 (2015). doi:10.5194/hess-19-341-2015
109. Tanoue, M., Hirabayashi, Y., Ikeuchi, H.: Global-scale river flood vulnerability in the last 50 years. Sci. Rep. **6**, 36021 (2016). doi:10.1038/srep36021
110. Touzi, R., Lopes, A., Bruniquel, J., Vachon, P.: Coherence estimation for SAR imagery. IEEE Trans. Geosci. Remote Sens. **37**(1), 135–149 (1999). doi:10.1109/36.739146
111. Townsend, P.A.: Mapping seasonal flooding in forested wetlands using multi-temporal Radarsat SAR. Photogramm. Eng. Remote Sens. **67**(7), 857–864 (2001)
112. Tsang, L., Kong, J.A., Ding, K.H.: Scattering of Electromagnetic Waves. Theories and Applications. Wiley, New York (2000)
113. Tsyganskaya, V., Martinis, S., Twele, A., Cao, W., Schmitt, A., Marzahn, P., Ludwig, R.: A fuzzy logic-based approach for the detection of flooded vegetation by means of synthetic aperture radar data. ISPRS Int. Arch. Photogramm. Remote Sens. Spat. Inf. Sci. **XLI-B7**(September), 371–378 (2016). doi:10.5194/isprsarchives-XLI-B7-371-2016
114. Twele, A., Cao, W., Plank, S., Martinis, S.: Sentinel-1-based flood mapping: a fully automated processing chain. Int. J. Remote Sens. **37**(13), 2990–3004 (2016). doi:10.1080/01431161.2016.1192304
115. Verdoliva, L., Gaetano, R., Ruello, G., Poggi, G.: Optical-driven nonlocal SAR despeckling. IEEE Geosci. Remote Sens. Lett. **12**(2), 314–318 (2015). doi:10.1109/LGRS.2014.2337515
116. Vosselman, G., Maas, H.: Airborne and Terrestrial Laser Scanning. Whittles Publishing, Dunbeath (2010)
117. Walker, J.P., Houser, P.R., Reichle, R.H.: New technologies require advances in hydrologic data assimilation. EOS Trans. Am. Geophys. Union **84**(49), 545 (2003). doi:10.1029/2003EO490002
118. Wang, Y., Hess, L.L., Filoso, S., Melack, J.M.: Understanding the radar backscattering from flooded and nonflooded Amazonian forests: results from canopy backscatter modeling. Remote Sens. Environ. **54**(3), 324–332 (1995). doi:10.1016/0034-4257(95)00140-9
119. Wang, Y., Imhoff, M.L.: Simulated and observed L-HH radar backscatter from tropical mangrove forests. Int. J. Remote Sens. **14**(15), 2819–2828 (1993). doi:10.1080/01431169308904311
120. Westerhoff, R.S., Kleuskens, M.P.H., Winsemius, H.C., Huizinga, H.J., Brakenridge, G.R., Bishop, C.: Automated global water mapping based on wide-swath orbital synthetic-aperture radar. Hydrol. Earth Syst. Sci. **17**(2), 651–663 (2013). doi:10.5194/hess-17-651-2013
121. Weydahl, D.J.: Flood monitoring in Norway using ERS-1 SAR images. In: IGARSS '96. 1996 International Geoscience and Remote Sensing Symposium, vol. 1, pp. 151–153. IEEE (1996). doi:10.1109/IGARSS.1996.516274
122. Zink, M., Bachmann, M., Brautigam, B., Fritz, T., Hajnsek, I., Moreira, A., Wessel, B., Krieger, G.: TanDEM-X: the new global DEM takes shape. IEEE Geosci. Remote Sens. Maga. **2**(2), 8–23 (2014). doi:10.1109/MGRS.2014.2318895

Remote Sensing as a Tool for Analysing Channel Dynamics and Geomorphic Effects of Floods

Margherita Righini and Nicola Surian

Abstract Over the past two decades, the use of optical remote sensing in fluvial geomorphology has become widely employed for several applications, due to improvements in geospatial technologies and data availability. However, applications focused on change detection of channel dynamics and geomorphic response to individual flood events are still relatively rare. Insights into the complexity of interactions driving geomorphic changes might be obtained by application of diverse remote sensing approaches, depending on several factors (e.g. temporal and spatial resolution, magnitude of detected change). An overview about remote sensing as a tool for channel dynamics and geomorphic response to flood detection is illustrated, including discussion about advantages and limitations of optical remote sensing.

Keywords Optical remote sensing • Geomorphic changes • Flood response

1 Introduction

Flood event studies require multidisciplinary approaches to enable the understanding of the phenomena [82]. Analysis of geomorphic responses to flood is a significant aspect of research, improving our capability of forecasting channel dynamics and related flood hazard [2, 12, 33, 46, 53, 56, 63, 64, 71, 80, 87]. Among the available tools, remotely sensed data have become a valuable source of information over the past two decades, and they are widely employed for several applications in fluvial geomorphology due to increasing geospatial technologies and data availability [54, 62].

This chapter aims to provide a review of the channel change detection analysis in response to flood events, with a specific focus on the use of optical remote sensing as a tool, often used in concert with other remotely sensed data. It specifically addresses river flood change detection analysis in the context of passive optical remote sensing, with some references to the active one. The main focus is on

M. Righini (✉) • N. Surian
Department of Geosciences, University of Padova, Padova, Italy
e-mail: Margherita.righini@studenti.unipd.it

A. Refice et al. (eds.), *Flood Monitoring through Remote Sensing*,
Springer Remote Sensing/Photogrammetry,
https://doi.org/10.1007/978-3-319-63959-8_2

channel dynamics and geomorphic effects of floods (e.g. bank erosion, channel aggradation, channel avulsion), while other processes related to floods, such as inundation processes, are not addressed in this work.

This chapter is organized into two main sections. The first section reports an overview about the advance in channel change detection analysis over time, strictly matching the progressive advance in technology and detection techniques. The second section describes the change detection analysis using optical remote sensing data by (1) analysing the potential for observing river changes through the use of multi-resolution and multi-temporal data; (2) discussing assumptions, advantages and limitations of optical remote sensing data applied to channel change detection; (3) reviewing some studies on river response to flood event, with a focus on the observed and investigated river changes; and (4), finally, describing how optical remote sensing data have been employed to support channel change detection during a flood event by means of an illustrative case study.

2 Channel Response to Floods: Detection and Technology Advancements over the Last Decades

Flood events may or may not affect channel morphology, shaping channels and floodplains. Since the early 1950s, research began to explore the effects of floods on riverine landscape, and the comprehension of their role in determining the physical processes driving channel forms and their impacts on morphodynamics became a central theme of fluvial geomorphology. The examination of riverine landscape revealed also to be strictly linked to advances in technology, and consequently it drastically improved over the past decades (Marcus and Fonstad [50]; Mertes [51]).

Leopold and Maddock [43] argued that channel-shape adjustment during individual floods occurred in response to a varying sediment load. In the early 1960s, research explored physical effects induced on the natural landscape by floods considering the relative importance of geomorphic processes of flood events with variable discharge [74, 87]. In 1963 Schumm and Lichty investigated periods of channel widening related to floods of high peak discharge and of floodplain construction related to floods of low peak discharge in a semiarid region. Stewart and La Marche [78] show that flood with recurrence interval >100 years largely determine valley morphology, channel pattern and location and the character of alluvial deposits. They mapped erosional and depositional features on post-flood aerial photographs and determined the amount of destruction of land and forests by comparing pre- and post-flood aerial photographs as support to field surveys. However, prior to the 1970s, the sensors mainly consisted of black and white photographic film, and the platform was an airplane in the context of geomorphological mapping [20]. In the 1970s, researchers began to analyse channel changes related to flood magnitude, focusing on the effects of floods of extremely rare occurrence [2, 24] and their effects across several climate regions [86]. However, most of these works were based on traditional field survey observations of river cross sections and relied on mostly qualitative, and rarely quantitative, flood-related impact studies [35].

In 1972 the launch of Landsat I started the first continuously acquired collection of space-based remote sensing data of terrestrial environments [85]. Nevertheless, in the early phases of satellite remote sensing, the data available had 80 m ground resolution, and the pioneering investigations in the field of application of remote sensing to flood were predominantly focused on flood monitoring and mitigation, with particular regard to the recognition of flood-prone regions of the USA [81] and flood extents [77]. Large pixel sizes, in many cases larger than the stream width, limited the use of satellite-based imagery in channel change detection and river feature mapping. Subsequently, and until the late 1990s, researcher have long used air photos to document changes in channel parameters, based on visual interpretation or photogrammetric analysis [27, 38, 50]. In the mid-1980s, imaging spectrometry became an earth observation tool, even if the increase of the spectral resolution required a lower spatial resolution, slacking space-borne hyperspectral development [17].

A significant shift in river research occurred during the mid-1990s, from field-based surveys of river cross section, coupled with qualitative interpretation of aerial photography through new tools available to fluvial geomorphologists [37]. The technical progress accomplished in the past two decades of research in remote sensing made a considerable contribution to river sciences [8] and, among these methods, optical remote sensing of rivers has experienced some of the more dramatic advances [50]. Remote sensing becomes a reasonable alternative and a good support to ground-based field survey when it can monitor or map the variable of interest [49]. In addition, digital elevation models (DEMs) became available during the 1990s [35]. Optical remote sensing and DEMs greatly improved the ability to create and manipulate detailed imagery of Earth's surface, resulting in opportunities to detect features not previously recognized [85]. Over recent years, additional advances in technology allowed to overcome some logistic and cost constrains in investigating changes in fluvial settings. The development of unmanned aerial systems (UASs), able to fly in nonurban areas at very low altitudes and to deliver very high spatial and temporal resolution imagery, reveals a significant potential in river sciences, offering very interesting alternatives to traditional platforms [8, 81, 88].

3 Channel Change Detection Analysis in Response to Floods

Change detection is the process of identifying differences and quantifying temporal effects of an object or a phenomenon using a multi-temporal data set [75]. Hence, the primary purpose of a change detection analysis is to characterize environmental change over time and space [84]. Information on river morphology and changes over time is commonly needed for water resource planning, river management [22] and hazard assessment and for a better understanding of the main causes of those processes [49]. Hence, the most efficient change detection techniques are able to quantify areal surface, direction and rate of change and to estimate the accuracy of change detection results [45]. Singh [75] reviewed a series of feasible procedures for change detection using remotely sensed data in environmental fields, identifying the post-classification comparison (i.e. comparison of independently classified images) as the most common method. The following issues are addressed in this section: (1) potential of observing river changes in response to a flood through optical remote sensing data; (2) the use of remotely sensed data for channel change detection across multi-resolution and multi-temporal data; (3) common assumptions, advantages and limitations of remote sensing techniques for channel change detection analysis; (4) the use of optical remote sensing data in concert with other remote sensing, topographic and geospatial data; and (5) the application of optical data for channel change detection by means of a real case study.

3.1 Potential of Observing River Changes Using Remote Sensing Data

An important prerequisite to assess a change detection related to a flood event is the identification, the qualitative or quantitative assessment, of riverine landscape forms and structures before and after the flood event. Optical remote sensing data allow change detection through several methods, ranging from classic visual skill-based interpretation procedures to semi-automated and automated methods based on imagery characteristics. Regardless of the used technique, the success of change detection from imagery will depend on both the nature of the change involved and the success of the image preprocessing and classification procedures (Bryant and Gilvear [6]). Most of the works documenting spatial and temporal variations of channel response to flood are mainly focused on two components of potential observing changes, such as planimetric 2D changes and 3D changes. Other contributions came up from studies illustrating hydraulic or geomorphic derived variables or indicators, which should develop a better understanding of the mechanisms responsible for channel changes to floods and further predictive models. We reviewed a number of works which focused on a wide spectrum of channel responses and magnitudes of floods (e.g. from moderate to extreme flood events), investigating channel or floodplain changes at different spatial and temporal scales, in diverse climate and geomorphic settings, using different remote sensing data sources and approaches (Table 1). Depending on the detection purpose, optical remote sensing data have been often used in concert with other topographic data integrating field surveys. Among the analysed works, some focused on channel planimetric 2D, while others on 3D changes, according to the dominant changes occurred and to data availability.

Table 1 Examples of application of remote sensing to river change detection analysis in response to floods

River	Flood event	Recurrence interval (year)	Detection purpose	Method	Remote sensing data	Spatial scale detection analysis	Time scale detection analysis	Reference
Planimetric 2D changes								
River Dee, Wales-England border, UK	n.a.	n.a.	Spatial and temporal trends in the channel planform	Pre- and post-flood aerial photographs comparison	Six sets of air photographs	Reach (18 km)	115 year period (1876–1992)	[25]
Eel River and its five tributaries, Northern California, USA	1955;	50 years	Variation in channel width and position	Pre- and post-flood aerial photographs comparison	Sets of aerial photographs at 3–7 year interval	Catchment	Three-flood events	[76]
	1964;	200 years						
	1997	12–15 years						
Sabie River, South Africa	February 2000	200 years	Changes in morphologic unit composition and spatial coverage	Aerial photographs comparison	Pre- and post-flood aerial photographs (1:10,000 scale)	Channel types (1–20 km); morphologic units (up to metre scale)	One-flood event	[26]
River Bollin, NW England	2000	n.a.	Occurrence and causes of multiple cutoffs on a meandering river	Aerial photographs mapping	Aerial photographs	Reach (600 m)	2000–2001	[27]
	2001							
Guadiana River, Spain	5–6 November 1997	>500-year	Sedimentary and erosive features	Pre- and post-flood aerial photograph mapping	Aerial photographs	Reach (500 m to 2 km)	One-flood event	[57]

(continued)

Table 1 (continued)

River	Flood event	Recurrence interval (years)	Detection purpose	Method	Remote sensing data	Spatial scale detection analysis	Time scale detection analysis	Reference
River Dane, NW England	> 25 years period	n.a.	Change in morpho-logical characteris-tics of meander bends	Photogrammetric mapping	Sets of aerial photographs (from 1984 to 2007)	Meander bends	25 years	[29]
Seven rivers along the Mediterranean coast, Israel	n.a.	n.a.	Geomorphic effect on river mouth	Pre- and post-flood aerial pho-tographs comparison	Pre- and post-flood aerial photographs	River mouth	1918–2005	[44]
Five alpine gravel bed rivers, western Austria	28 August 2005	500 years	Variation in channel width	Pre- and post-flood orthoimage comparison	Pre- and post-flood aerial photos; LiDAR data (1 × 1 m)	Reach (range: 20–56 km); sub-reach (500–5000 m); cross section (200 m spacing)	One-flood event	[33]
Rapti River, India	2011 monsoon floods	>500 years	Channel migration	Aerial photograph comparison	SRTM DEM data (90 m); satellite images (Landsat MSS 60 m, TM 30 m, ETM+ panchromatic-8 band, 15 m)	Reach	Monsoon season-flood events	[34]
White and Saxtons Rivers, Vermont, USA	Tropical storm Irene 2011	>100 years	Channel geometry changes	Aerial photograph comparison	Pre- and post-flood colour aerial imagery (1 m); WorldView-1 panchromatic satellite imagery (0.50 m); DEMs	Reach	One-flood event	[7, 46]
Six mountain tributaries, Italy	25 October 2011	>100 years	Variation in channel width	Pre- and post-flood aerial photo comparison	Aerial photos (0.50 m)	Reach (1–6 km); sub-reach (300–500 m)	One-flood event	[80]

3D changes								
River Tay, Scotland, UK	January 1993	40–65 years	Land cover change; bathymetric change	Classification comparison and bathymetric mapping	Airborne imagery-Daedalus 1268 (ATM); colour aerial photography of 1:12,000	Reach	1992–1994	[6]
Lockyer Creek, South East Queensland (SEQ), Australia	January 2011	2000 years in the upper catchment to 30 years downstream	River bank erosion	DoD	Multi-temporal LiDAR; high-resolution pre- and post-flood aerial imagery	Catchment	One-flood event	[23]
Rio Puerco, New Mexico, USA	2006	n.a.	Topographic changes	DoD	QuickBird II panchromatic image (0.6 m); multiple airborne LiDAR (0.5 m)-derived DEMs	Segment (12 km)	One-flood event	[60]
Elbow River, southwestern Alberta, USA	June 19 and 23 2013	500 years	Sediment erosion and deposition	DoD; geostatistical models (pre- and post-flood variograms)	UAS survey: orthoimagery; DEMs (5 cm/pixel and 4 cm/pixel)	Reach	One- flood event	[81]
Planimetric and 3D changes								
Lower Yuba River, Sacramento Valley, California, USA	Eight major floods (1907, 1909, 1955, 1964, 1970, 1986, 1997, 2005)	n.a.	Planimetric changes; volumetric changes in sediment storage	Pre- and post-flood aerial photo comparison; DoD	Pre- and post-flood aerial photographs; DEMs (3 × 3m)	Reach	1906–1999	[22]

(continued)

Table 1 (continued)

River	Flood event	Recurrence interval (year)	Detection purpose	Method	Remote sensing data	Spatial scale detection analysis	Time scale detection analysis	Reference
River Dane, NW England	>20 year period	From 9 to 57 years	Rates and occurrence of erosion, deposition, calibre of sediment, and morphological changes	Photogrammetric mapping to an accuracy of ±0.5 m	Set of aerial photographs	Reach (10 km)	1981–2002	[29]
Thinhope Burn, Northern England, UK	17 July 2007	n.a.	Spatial patterns of erosion and deposition; volumetric changes	Pre- and post-flood aerial photo comparison; DoD	Pre- and post- flood aerial imagery; DEMs	Catchment and reach	One-flood event	[52]
Lockyer Creek, South East Queensland (SEQ), Australia	January 2011	2000 years in the upper catchment to 30 years downstream	Areal changes extend of geomorphic features, variation in channel width	Pre- and post-flood aerial photos comparison; DoD	Multi-temporal LiDAR-derived DEMs (5 cm); pre-and post-flood aerial photos	Reach	One-flood event	[83]
Magra River, Italy	25 October 2011	From 30 to 200 years	Variation in channel width; eroded volumes of sediment	Pre- and post-flood aerial photo comparison; DoD	Pre- and post-flood aerial photos (0.50 m); LiDAR derived DEM	Reach (1–6 km) sub-reach (300–500 m)	One-flood event	[56]

n.a. not available data

3.1.1 Planimetric 2D Changes

Channel planform change has been the research focus of many fluvial geomorphologists (Gilvear and [17]). The main intent of planimetric change detection analysis is to characterize or quantify channel and floodplain planform changes [68]. Airborne and satellite-based remote sensing data have long been used to map channel boundaries, bars, floodplain cover, in-stream islands and other features [20]. When a series of several images or at least a set of two images is available, it is possible to observe changes on different timescales [21], related to one or to a series of flood events with diverse magnitude and recurrence intervals.

Most of the studies analysing planimetric changes in response to diverse event magnitude (i.e. from extreme to moderate floods) used airborne imagery across diverse spatial scales (i.e. from catchment to morphologic units metre scale) and time scales (i.e. from one flood event to wide temporal intervals >100 years). These studies mainly focused on (1) channel width changes [25, 33, 56, 76, 80, 83], (2) channel and geomorphic features pattern changes [27, 29, 44, 52, 57, 82], (3) substrate type changes [26] (Fig. 1), (4) channel shifting [76] (Fig. 2), (5) bank retreat [23, 56] (Fig. 3) and (6) land coverage changes [6].

Studies of large rivers are generally based on satellite imagery which provide spatially extensive images over large areas for low costs. Therefore, if the pixel size is suitable for the river size, the regular revisit frequency of orbital sensors allows easier multi-temporal approach analysis compared to airborne data.

Kumar et al. [34] used multi-temporal satellite images to assess the impacts of 2011 monsoon flood in the alluvial Rapti River (India) quantifying both areal (i.e. channel area, channel belt area, bar area) and linear characteristics of channel belt before and after the flood (i.e. active channel and channel belt width, sinuosity, channel central line shifting). Authors used Landsat MSS, TM, ETM+ data, with spatial resolution of 60, 30 and 15 m, respectively, in concert with SRTM elevation data to map geomorphological features, quantify geomorphic processes and resultant landforms at various spatial scales ranging from basin to reach level. In this case, employment of medium spatial resolution data turned out to be suitable for channel having an average initial channel width greater than 60 m at reach scale and notable magnitude of change in terms of channel lateral expansion and shifting and channel belt area variations. Higher spatial resolution space-borne imagery were used by Buraas et al. [7] and Magilligan et al. [46] to investigate channel widening that occurred in the Williams, White and Saxtons Rivers, Vermont (USA) during the Tropical Storm Irene. Post-flood panchromatic satellite imagery from WorldView-1 (0.50 m of spatial resolution) were used to define major erosional features across floodplains\ and channel width changes exceeding the natural "range of variability" of channel width observed in the absence of extreme floods (i.e. pre-Irene 20-year period) [46]. The available imagery allowed to analyse large portions of affected channels (i.e. up to 40 km) at reach scale showing a channel widening up to 12 m of net change, from average pre-flood width of 18 and 48 m, respectively, for the Saxtons and White Rivers (Fig. 4). Furthermore, Buraas et al. [7] used satellite

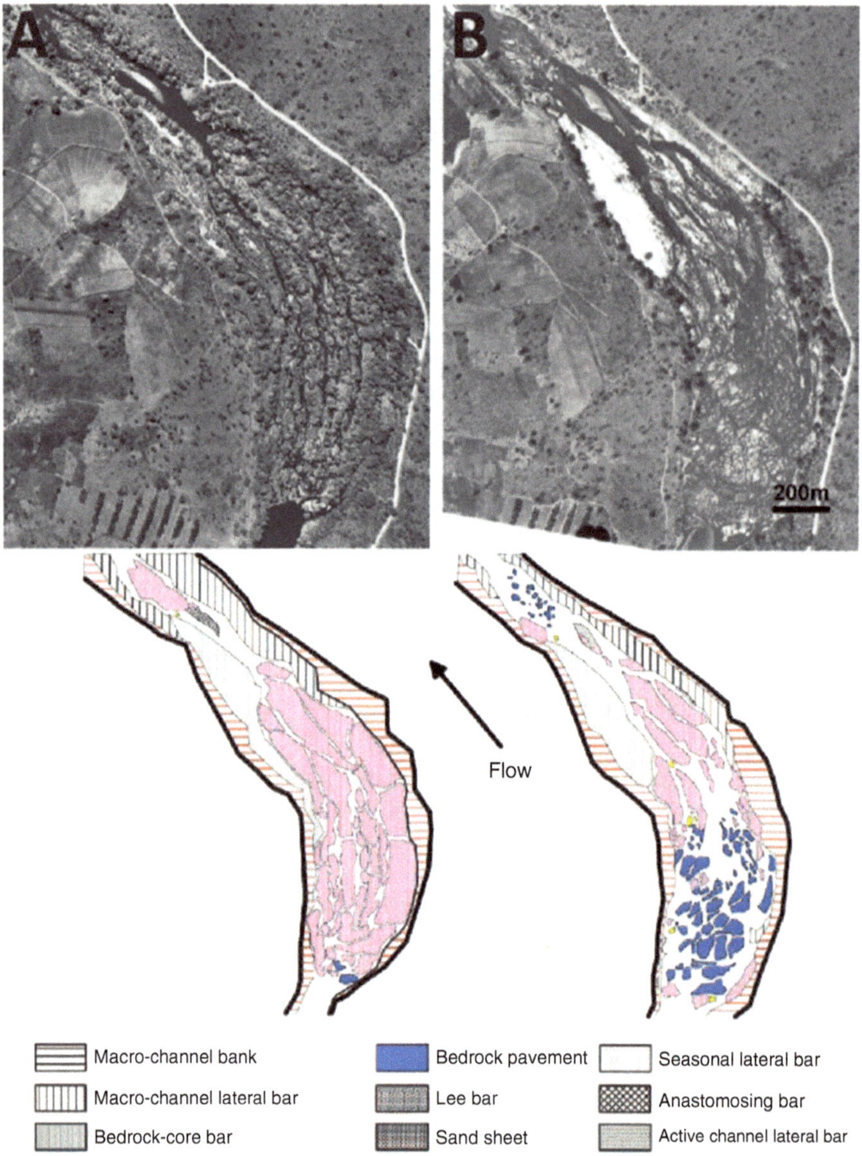

Fig. 1 Example of cohesive mixed anastomosed channel type on the Sabie River, showing (**a**) pre-flood and (**b**) change in morphologic unit composition following the extreme flood of 2000 on aerial photographs (1:10,000 scale) [26]

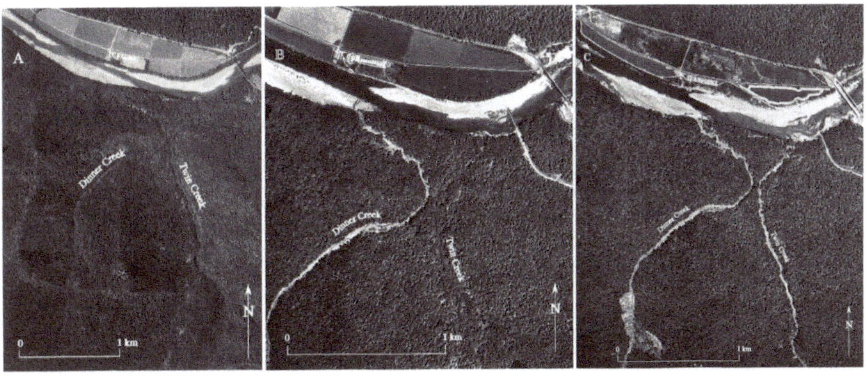

Fig. 2 Variation in channel width and position on a set of aerial photos of Dinner and Twin Creeks: (**a**) before the 1955 flood (1954 photograph), (**b**) after the 1955 flood (1960 photograph), (**c**) after the 1964 flood (1966 photograph). Hillslope failures were also identified on the 1966 photos in both basins [76]

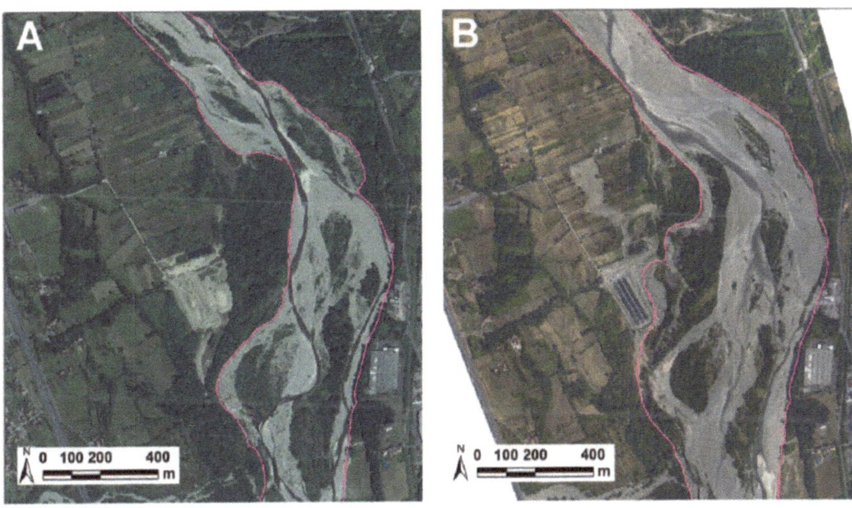

Fig. 3 Digitalization of the active channel on colour aerial photos (spatial resolution = 0.50 m) to evaluate variation in channel width (**a**) before the flood and (**b**) after the flood [56]

data as support for developing a method able to predict more susceptible reaches to major geomorphic change. Space-borne data were also employed as support in assessing the distribution and the role of riparian vegetation (i.e. type and density) in influencing geomorphic effects during extreme flood affected the Rio Puerco, New Mexico (USA) in 2006 [60]. Principally QuickBird II panchromatic images (0.61 m of spatial resolution) were employed to map the distribution and pattern of plants in the landscape using unsupervised classification approach in order to categorize each pixel as either bare ground or vegetation (i.e. darkest areas). Satellite imagery turned out to be effective for identifying young plants and sandbar willow compared

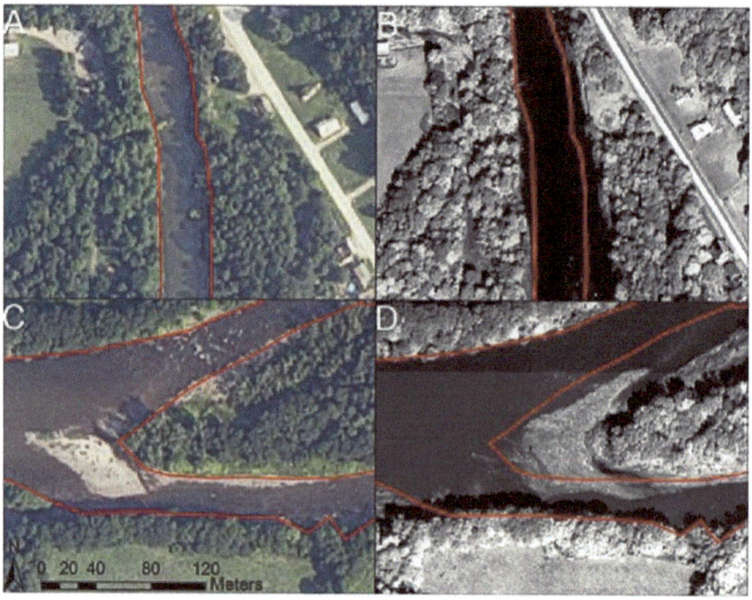

Fig. 4 Width changes on the White River on colour aerial photos (spatial resolution = 1 m): (**a**) pre-image with channel delineation, (**b**) post-image displaying bank erosion, (**c**) pre-image with island delineation, (**d**) post-image displaying erosion of the head on an island as cause for channel widening [7]

to airborne LiDAR which was not suitable for detecting thin vertical stems and branches, because of their difficulty in reflecting laser signals.

3.1.2 3D Changes

Recent methodological advances, including Light Detection and Ranging (LiDAR), Structure from motion (SfM) and terrestrial laser scanning (TLS), allow investigation of other aspects of channel adjustment to floods, such as the investigation of the three-dimensional river structure and changes (e.g. scour and fill, bank erosion) [81]. Bed-level change, lateral erosion, variation in sediment storage, floodplain scour, fill and associated volumes of erosion and sedimentation are the more common 3D channel changes detected as response to a flood. However, one of the main issues related to 3D channel changes detection analysis is the diverse approach needed for exposed (e.g. channel bars or exposed channel beds) and submerged areas. Aerial photogrammetry and laser altimetry are employed to detect and quantify channel changes for exposed areas [38], while optical and multi- and hyperspectral imagery might detect variations in water depth in shallow non-turbid and deeper and more turbid channels, respectively [20, 41, 58]. The "DEM of Difference" (DoD) approach, derived by photogrammetry or LiDAR, is one of the most common methods to analyse 3D river changes (Table 1).

Hooke [28] recorded the magnitude and conditions of process and morphological response of the channel (i.e. rates and occurrence of erosion, deposition, calibre of sediment and morphological changes) related to discharge characteristics variations for a sequence of more than 50 bends in a 10 km length reach in an active meandering river in NW England over a 20-year period. Measurements and monitoring of the study reach consisted of detailed field mapping, bends ground photographs comparison and three sets of aerial photographs. Bank lines were mapped photogrametrically to an accuracy of ±0.5 m from air photos and the digitized courses compared in ArcGIS to produce calculations of erosional and depositional areas and rates [48].

Besides mapped riparian vegetation type and distribution, Perignon et al. [60] presented the results of an airborne LiDAR differencing study (0.5 m of spatial resolution) of the topographic effects on the Lower Rio Puerco, New Mexico (USA). Authors analysed erosion and deposition volume changes along the channel and the floodplain and removal of vegetation from river corridor.

Some works integrated the planimetric change detection analysis with topographic analysis. Grove et al. [23] and Thompson et al. [82] carried out a multi-temporal LiDAR-derived DoD analysis to map and quantify the three-dimensional form or morphology, volume and rates of different of bank erosion processes before and after the flood (Fig. 5). The DoD approach was also used to quantify the eroded volumes of sediment derived from riverbank retreat [56], the spatial patterns of erosion and deposition and volumetric changes [52], volumetric changes in sediment storage [22]. Tamminga et al. [81] used the same approach employing digital elevation models produced from photogrammetry using pre- and post-flood high-resolution (4–5 cm/pixel) imagery to detect reach-scale spatial patterns, volume and area of erosion and deposition acquired by unmanned aerial systems (UASs) in Elbow River, Alberta (USA). They also applied a different approach to investigate reach-scale morphology and changes applying variogram analysis to topographic data elevations in order to determine characteristic horizontal and vertical length scales and variance of elevations taking into account varying channel complexity and anisotropy in different directions.

3.1.3 Derived Variables and Parameters

Research is moving to use optical remote sensing data, usually in synergy with other remotely sensed (i.e. LiDAR or SAR) and topographic data (e.g. DEMs, DTMs), to extract other variables and parameters or indicators, potentially explanatory for understanding river response to flood and for developing predictive models. These approaches enable to extract and map the following variables or parameters: (1) hydraulic variables, such as water surface slope, stream power [31, 48] or specific stream power and shear stress [3, 16, 46], (2) morphological parameters and indices, such as confinement index or confinement degree, defining valley setting [69, 71, 80], the sinuosity index, the anabranching index [79], the braiding index [1], the alluviation index [26] and the width ratio [33]; (3) channel instability

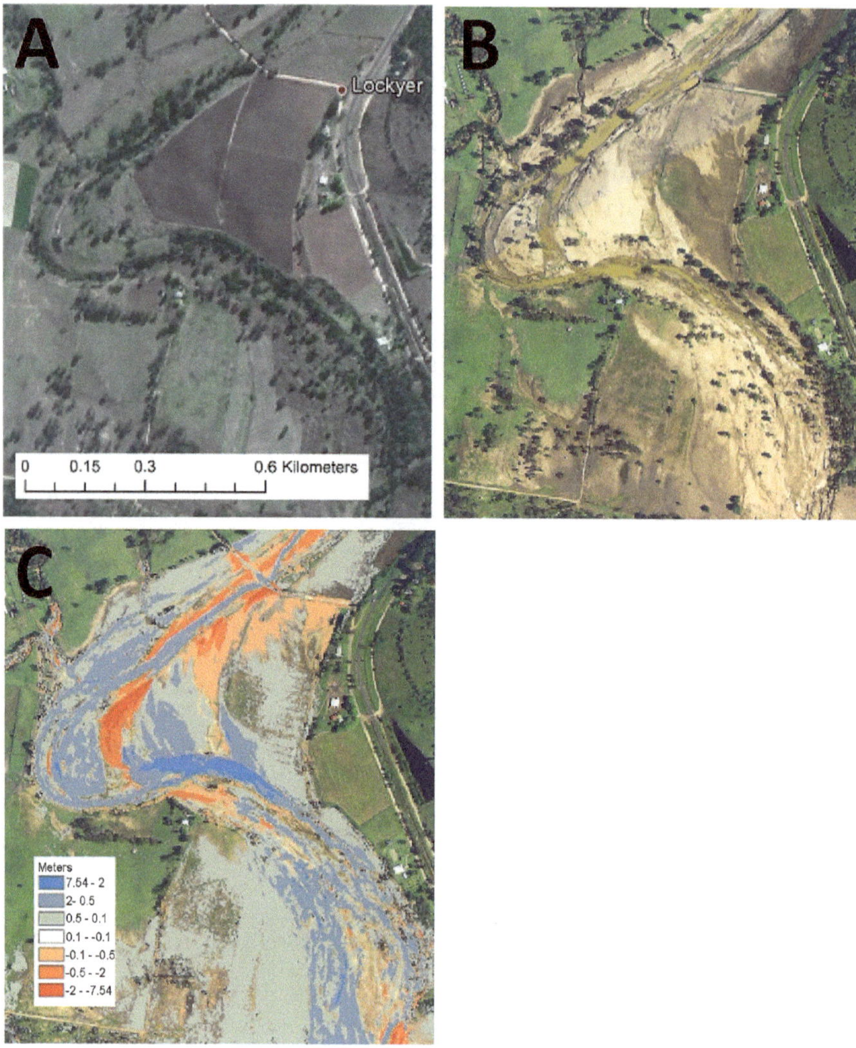

Fig. 5 Change (i.e. erosion/deposition, variation in channel width and reorganization of the channel morphology) along unconfined reach of Lockyer Creek showing (**a**) pre-flood aerial photos, (**b**) post-flood aerial photos, (**c**) elevation change from DoD [83]

or susceptibility parameters, such as the mobility index [4] or the erosion and deposition index [28, 61] and the bend stress parameter [7]; and (4) sediment connectivity indicators, like the connectivity index [10].

3.2 *Flood Effect Detection Across Spatial and Time Scale Using Optical Remote Sensing Data*

Natural processes frequently operate on larger spatial scale and longer time scales than traditional river sciences and management [8]. In the past, understanding of environmental changes was constrained by inability to quantify spatial variation and change in fluvial landforms over long distance or long time periods. The result of using numerous local measurements, or sporadic, spatially widespread measurements, led to consider the channel as a discontinuous system [50]. The relevance of such small scale for developing mechanistic models of changing river channels and understanding larger systems might be problematic, and a generalization from discrete patches to an integrated whole might be questionable [89]. Then again, both the limited temporal scale of flooding and the transient effects of changes on the river channel and floodplain, or the low frequency of geomorphic effective flood events, constrained forecasting and prediction perspectives. Access to and availability of remote sensing data, operating at a range of spatial and temporal scales, enhanced the ability to survey wider areas, particularly in large or inaccessible fluvial systems, and allowed change-over-time analysis [18], overcoming the issue of channel response to floods as temporary phenomena, usually followed by a variable recovery period [86]. Indeed, remotely sensed data and imagery is the only approach that could conceivably give continuous data over entire catchments [18] and provide synoptic information [19, 50], retrospective viewing and multi-temporal analysis.

Understanding of river processes and morphodynamics requires investigations at larger scales considering a river catchment as a holistic system, involving both floodplain and hillslope dynamics, and land use and human impact aspects. Remote sensing approaches provide the possibility to investigate from catchment to local scale and determine changes affecting the entire basin, such as disturbances on channel-hillslope coupling, in order to better understand process linkages between fluvial and hillslope systems [14, 71, 76, 80]. Moreover, the increasing availability of metre-to-centimetre resolution imagery enables the exploration of smaller scales and in-channel morphological changes and in-stream parameters. Recent research indicates that optical imagery can provide fine-resolution watershed coverage of in-stream features, representing rivers as continuous systems [50]. Hence, the necessity to investigate rivers and their changes across different spatial and time scales has become more challenging, and the data acquisition should depend on both the nature and the scale of the purposed investigation [65].

Two interesting frameworks that link spatial scales, temporal scales and data sources have been proposed by Prienstall and Aplin [65] and Rinaldi et al. [71]. Prienstall and Aplin [65] have suggested suitable spatial scales and data sources for river environment representation in digital form, taking in consideration different study scales (i.e. from catchment to micro-topography) (Fig. 6a). Rinaldi et al. [71] have proposed an integrated approach for investigating geomorphic response to extreme events, highlighting the complexity of interactions driving change variability in riverine landscape across several spatial scales. Such approach outlines a summary of the spatial scales and methods related to the various components of an overall analysis of the geomorphic response to a flood event (Fig. 6b).

3.3 Assumptions, Advantages and Limitations

In this section, we summarize some of the advantages and limitations of employing optical remote sensing data in riverine landscape studies. Remote sensing compared to traditional data collection (i.e. cartographic and field-based data) reveals several advantages [20], though it might encounter some issues and limitations, and specific assumptions are needed to be correctly employed. First, remote sensing provides an approach for assessing the landscape as a function of the spatial, spectral, temporal and radiometric resolutions at which sensor-platform systems operate. These four remote sensing resolutions combine to characterize the landscape from local to global spatial scale [84]. Spatial resolution, image extent and spectral characteristics play a large role in determining whether or not a particular sensor or data is capable of detecting individual features [32]. However, the most noticeable optical constraint is that the stream must be visible from above, and consequently optical remote sensing of streams cannot be done where trees, bridges, woods or other obstacles overhang the channels. The no-obstruction criterion is especially constraining in headwater streams or along stream banks in dense forested environments [50]. Magilligan et al. [46], for instance, interpolated the channel boundary, at locations closest to the channel edge, to deal with the channel banks covered by shadows or overhanging vegetation.

Spatial coverage and temporal resolution In the case of very large rivers, using aerial or space-borne remote systems is the only way to observe and quantify the overall morphology of the river [20]. According to the previous section, the use of remote sensing for tracking flow variability and flood effects is particularly important in areas without monitoring systems or where access is limited. Indeed, one of the main recognized advantages of using remote sensing is the global view [8, 51, 82]. In addition, temporal resolution and coverage provided by remote sensing systems is crucial in change detection studies [8]. Temporal resolution refers to the elapsed time between consecutive imagery. It turns out to be important in order to collect consecutive measurements and to carry out multi-temporal analysis and historical perspectives. Therefore, the temporal resolution has to be finer for acquisition of short-lived flood events [51]. Whereas studies of large rivers based on satellite imagery have been able to exploit the regular revisit frequency of orbital sensors, airborne data is not acquired with the same regularity [8]. However, commissioned flights are often necessary to attain suitable temporal resolution, particularly for a short-term change resulting from a single flood event (Gilvear and Fonstand 2003). More recently, the acquisition of hyperspatial imagery (i.e. imagery with sub-decimetric spatial resolutions) [9, 67] at monthly or even daily

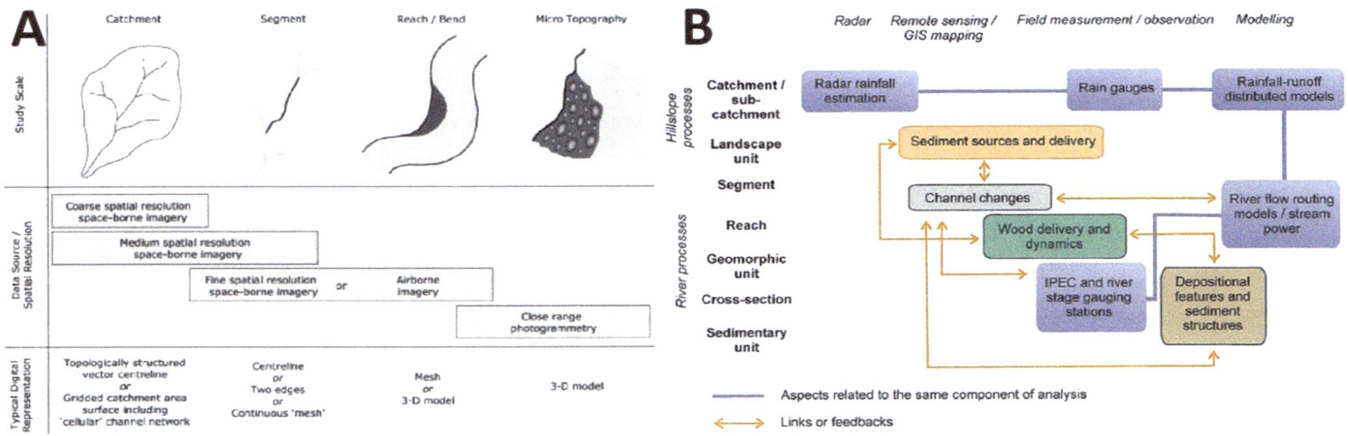

Fig. 6 Spatial scales and approaches for analysing river environments and investigating geomorphic response to extreme events proposed by (**a**) Priestnall and Aplin [65] and (**b**) Rinaldi et al. [71] respectively

temporal resolutions is becoming logistically and economically possible for river reaches of hectametric or kilometric scales. An essential assumption to apply the optical remotely sensed approach in river change detection analysis is the temporal proximity of images to the flood event that must be as much as possible representative of the pre- and post-flood geomorphic characteristics, avoiding the error of detecting changes not produced by the latest flood. Buraas et al. [7] and Magilligan et al. [46] deal with changes in channel width in the absence of extreme floods, determining thresholds of variations over the 20 years preceding the Tropical Storm Irene, to avoid errors related to temporal resolution of data. In the study of Lichter and Klein [44], the temporal resolution of the observations was limited to historical aerial photographs, which represent a specific moment in time. However, the database used in that study enabled an exceptional outlook into past flood events, covering an 85-year period. Recent advance in technology allows the widespread availability of unmanned aerial system (UAS), existing in a very wide range of sizes and purposes, and available on the civilian commercial market [8].

Radiometric resolution Describes the sensor's ability to detect small changes in radiance and depends on the manner in which the continuous upwelling radiance signal is converted to discrete, digital image data. For example, an investigation of channel change might benefit from a highly sensitive detector with 12-bit radiometric resolution that enables very precise water depth estimates [42].

Spectral resolution Spectral resolution refers to the number of wavelength bands in which radiance measurements are made, as well as the location and width of these bands [42]. The ability to accurately detect features from remote sensing therefore depends not only on increasing the number of bands beyond the visible spectrum but also on an improvement of the spectral resolution. Indeed, the knowledge of spectral characteristics of the feature of interest and their response to environmental variables may enhance the use of remotely sensed data [20]. In image interpretation analysis, the human eye tends to assume contextual information for feature identification and discrimination, whereas remotely sensed data allow distinguishing spectral discontinuities [65]. Panchromatic images, having a single, wide band and very high resolution, are generally used for detail rendering, while multi-spectral images having several bands and lower resolution may have different reflective properties at different wavelengths. The greyscale data of panchromatic images can be used to map channel planform, lateral migration [59] and, to a lesser extent, variations in water depth. However, incorporating spectral information allows channel morphology to be measured with greater confidence and the possibility of examining other river attributes [42]. Spectral resolution can be crucial in identifying certain materials, such as chlorophyll, based on their reflection of light as a function of the wavelength of the incident light. Therefore, in the case of limited spatial resolution, the availability of spectral bands achieving greatest contrast between land and water is most suitable for channel planform detection [20]. Many airborne sensors and commercial satellites provide multispectral data, most often consisting of blue, green, red and near-infrared bands, and more advanced hyperspectral instruments measure radiance in a larger spectrum. For instance, the recently launched WorldView-3 satellite proposes a marked improvement in spectral terms with eight bands in the visible and near-infrared range and eight bands in the shortwave infrared (SWIR). Classification of imagery into surface categories based on their spectral properties may be carried out by several methods, allowing to support or speed up visual interpretation. Human features and structure, or riparian vegetation, might influence or limit significantly the behaviour of channel response to flood events. These features often stand out clearly on remotely sensed images, because of their shape and texture and their different composition relative to surrounding materials, which enables their detection through automated, spectrally based techniques [49]. Bryant and Gilvear [6] illustrated the potential of multispectral and hyperspectral airborne remote sensing (i.e. ATM) for detecting and quantifying changes in both fluvial submerged and exposed landforms (e.g. bar head accretion, bar tail formation and extension, bar dissection, localized bank erosion) and riparian land cover in a wandering gravel bed river before and after a flood. A maximum likelihood classifier was used on each image, and change detection analysis was carried out using a classification comparison approach. Bathymetric mapping was undertaken by applying a specific algorithm to ATM data.

Spatial resolution Spatial resolution defines the size of the smallest object which can be resolved on the ground [8], and it is just one of many variables that influence the extraction of patterns and the study of processes using remotely sensed imagery [65]. The difficulties in extracting river forms from imagery relate in part to the "precision" with which one can identify and map a river feature, as determined by the spatial resolution offered by the instrument [65]. The identification of a feature might be difficult because of pixel size larger than the object, which may imply the possibility of mixed pixels containing a number of several objects (e.g. bed material, water, vegetation) and hence potentially misinterpretation [20]. Carbonneau and Piégay [8] argued that there is no absolute rule for the number of pixels required to define a simple object and identified a minimum of 5×5 pixels required to get an approximation of the object shape, while 3×3 or even 2×2 pixels are required to establish the presence of an object of undefined shape in the image. However, users should balance the potential inaccuracies of remote sensing-based mapping techniques with the pixel resolution, and the selection of an appropriate spatial resolution should require knowledge of the stream of interest and a thoughtful evaluation of the purpose of the study [40]. Therefore, a pixel size that is adequate for one reach might not be suitable for other reaches. For instance, Ortega et al. [56] mapped morpho-sedimentary features in each reach on a 1:4500 scale topographic map, showing only features large enough to be mappable at that scale. Hyperspectral data also create opportunities to apply more sophisticated approaches such as spectral mixture analysis, which could be especially helpful along the banks. Therefore, the purpose of an investigation may determine whether imageries are appropriate for the extraction of distinct objects or features [65].

River size One of the main critical factors in channel change or feature detection analysis is the relationship between the width of the river and the spatial resolution of the image [65]. The size of the stream relative to the sensor's ground instantaneous field of view can largely determine the utility of remote sensing for mapping channel morphology and in-stream features. For systems with coarser spatial resolution, pixels might be contaminated by radiance from the stream banks, and a pixel size of one half the mean channel width is a basic minimum requirement [40]. Certain river settings do not work well for optical remote sensing, in particular high energy and small headwater streams. Investigation of relatively small channels may be also complicated by atmospheric conditions, vegetation cover and associated shadows along banks, making more complex feature digitization or extraction [40, 65].

Magnitude of change and accuracy Many authors recognized that estimates of channel changes are often affected by several errors [71]. Positional inaccuracy in digitized features considering channel or floodplain changes may often originate a spatial error being similar or greater than the magnitude of geomorphic changes [65]. Some works seemed to rely on inappropriate assumptions, where the magnitude of the error estimated using the most conservative methods is close to the magnitude of the measured change [55]. Mount and Louis [55] argued that rates of river channel migration or widening are only valid if it can be demonstrated that the amount of change in the measured parameters from images exceeds the measurement errors. They concluded that estimation and propagation of error in measurements of river channel movement from aerial imagery requires knowledge of both the systematic error components and the random error components associated with the precision of feature identification and that it is necessary to consider the following issues: (1) a number of possible assumptions relating to the homogeneity/heterogeneity of error components, directional and temporal independence of errors and (2) the magnitude of systematic error compared to random error components. Recent advances in combining the output of global positioning system with image capture has increased the potential for accurate identification of an absolute location and allowed ground records to be matched to individual pixels on imagery, permitting more accurate image calibration and validation [20]. Changes may be the result of cumulative errors. Scale distortions in the original imagery, poor selection of ground control points used for georectification and the algorithms used to transform the image to a specific coordinate system are possible sources of errors. These errors can be considerable and may sum up when attempting to detect change over time [30]. Hence, analyses of channel changes should include error assessment.

The majority of digital change detection techniques depend upon the precise accuracy of geometric registration of two images, which is very difficult to achieve due to the lack of accurate ground control points [75], and on different scales between the pixel dimension and the object to be analysed [51]. In spite of the need to distinguish true geomorphic change from errors associated with image registration, no widely accepted framework for defining positional errors in image time series has been established [39]. Standard methods of characterizing accuracy in remote sensing assume that the ground data are correct [49]. Users should also consider the potential inaccuracy of supervised and unsupervised imagery classification techniques, though standard methods of accuracy assessment are normally included [51]. However, spatial errors and uncertainties related to geomorphological change digitization and quantification, and the definition of a rigorous protocol for error assessment remain a controversial issue for many authors. Marcus and Fonstad [50] highlighted the lack in bibliography about what levels of accuracy and precision are required to answer specific research questions, preventing a wider use of remote sensing techniques in fluvial geomorphology. In response to this lack of error quantification, Mount and Louis [55] provided methods for estimating and propagating error in bankfull width measurements made from aerial photographs in a GIS. The first issue, which came up from this latter study, is given by the multiple possible definitions of bankfull. They assumed that it is appropriate to locate bankfull from aerial photos according to the location of boundary features and the water-sediment interface, where overhanging vegetation does not obscure banks. Sloan et al. [76] determined the minimum size of detected features based on the resolution of the aerial photographs and accuracy of measurements collected in the field, obtaining a 7% of error and determining a conservative estimation of change in channel width. Gurnell [25] adopted several strategies in order to minimize errors in extracting information on channel change from air photographs such as (1) a standard definition of river bank location using channel bank vegetation limit, (2) a standard set of spatial locations from which multi-temporal comparisons are made (i.e. cross sections), (3) a definition of a set of standard control points (i.e. from 15 to 25 points for individual photographs), (4) the employment of air photograph coverage of similar scale, (5) the visual interpretation provided by the same operator and (6) the use of standard non-linear least-squares transformation to minimize horizontal distortions. Buraas et al. [7] and Magilligan et al. [46] have considered several thresholds for defining a significant change in width, including one to four times the standard deviation in width changes in the absence of extreme floods. These thresholds are used to provide natural changes in width (i.e. not due to extreme flooding) and to account for the apparent variation due to different sources of error (e.g. shading, resolution of the imagery, overhanging vegetation covering the edge of the channel and operator error). Ghoshal et al. [22] pointed out that different flow stages at dates of aerial photograph acquisitions might introduce erroneous planimetric indications of bar erosion or deposition because high flows on a later date may give a false measure of bar erosion, while low flows on a later date may misleadingly suggest bar deposition. However, in the case of high-magnitude channel changes, some authors considered negligible errors related to georectification and digitizing [80]. More recently, Lea and Legleiter [39] introduced a framework based on a leave-one-out cross-validation (LOOCV) approach to characterize the spatial distribution of image registration errors in the analysis of channel change for five sequential 1-m pixels aerial photographs pairs of Savery Creek, Wyoming, USA (i.e. from 1980 to 2012). Their findings revealed that spatially distributed estimates of registration error enabled detection of a greater number of statistically significant lateral migration vectors rather than the standard metrics, such as spatially uniform RMSE or 90th percentile of GCP (i.e. ground control point) error [39].

The assessment of the error for 3D channel change detection approaches appears to be clearer. For instance, the uncertainty related to the DEM of Difference approach required the assessment of the minimum level of detection (LoD) that is detectable above the noise of the data [5, 81]. Therefore, the uncertainty might be either spatially uniform, constant within zones of similar characteristics, or variable for every cell of the raster [36, 60]. However, the uncertainties related to this approach might be related to several sources of error, such as systematic error affecting the accuracy of the measurements, random error affecting the precision of the data [23, 83], interpolation errors arising during the creation of DTMs from point data or filtering errors referred to discarding all off-terrain points in order to reconstruct the bare-earth only [13, 52, 60, 81].

Field validation and data matching As argued in the previous sections, satellites offer to landscape studies a vantage point of Earth observation, computer compatibility of sensed data, historical perspectives and near-global coverage. Nevertheless, spatial- and temporal-scale limitations of remotely sensed data imply that in-depth fieldwork, using both traditional and technologically advanced data collection techniques, will continue to be an integral part of the science of geomorphology [84]. Many researchers have found remotely sensed data a valuable tool for quantifying and detecting channel features and changes; however, few of them have addressed the problems of quantification and validation of change within the river and floodplain system from multi-date airborne or space-borne imagery [6]. The comparison between manifold field measurements and remote sensing techniques reveals a lower accuracy of remotely sensed data [50]. The dissimilarity between remote sensing data and field information is a general problem because of the mismatch between the scale of the objects and the spatial resolution of the data [51]. Often a direct comparison with field records is employed; however, advanced error analysis typically lacks due to sparse distribution of field data [51]. Ghoshal et al. [22], in their study involving aerial photographs rectification processes for planimetric analysis of lateral channel migration, used ground control points collected in the field near river channels, to ensure greatest accuracy in rectification where most channel-change measurements were taken. On the other hand, the change detection algorithm employed by Bryant and Gilver [6] to ATM data revealed subtle changes not noticed in the field. A remaining challenge is to achieve an exact match between the categories of landscape classification from the remote sensing analysis and data from the field survey or model outputs [51]. For Marcus and Fonstand [50], logistical obstacles concern ground validation procedures on time of imagery acquisition, which is difficult to accomplish at watershed extents, particularly in large and difficult to access river basins.

Costs and data availability The spatial extent of the study and the number of measurements needed over time are factors strictly driving costs [49]. Remote sensing becomes a reasonable alternative to ground-based field surveys when it can monitor or map the variables of interest and when it can do so on a cost-effective or safer basis than ground-based techniques. One advantage of optical imagery for remote sensing of rivers is that these data are widely available through government mapping agencies and commercial vendors that usually collect and archive airborne and satellite optical imagery also providing historical records. Aerial surveys are not cheap to commission, and repeat coverage at short time intervals is usually limited or impossible [11]. On the other hand, imagery from satellites is relatively cheap or freely available, and the scale of the events roughly matches the resolution of the satellite imagery [32].

3.4 Combination of Optical Techniques and Other Geospatial Data in Channel Response to Flood Detection

In many studies, the use of combining diverse data sources and processing made the channel change detection analysis more robust and accurate (Table 1) [47]. The widespread availability of geographic information systems (GIS) has allowed the wide use of optical imagery for channel change mapping, because it made easier the digitization of river features from aerial or satellite imagery and the assessment of a wide variety of spatial and change-over-time analyses [49]. The two technologies might be joined for a better examination of the landscape, investigating the interrelationships of scale, pattern and process [71, 84]. Specifically, GIS technology offers an analytical framework for data synthesis, capture, storage, management, integration, analysis, display and forecast [73, 82, 83, 88]. In the case of geomorphic changes in flood detection analysis, GIS techniques allow to make several operations (e.g. overlay, intersect, merging) enhancing a multi-spatial and temporal analysis and to characterize and quantify channel changes [15]. Particularly, the possibility of digitizing channel features allows the assessment of the pre-flood morphological state of the channel and the definition of the post-flood channel morphological state in a qualitative and quantitative way.

The merging of optical remotely sensed data with other spatial information within a GIS offers the possibility of measurement and quantification of channel changes by using different data sources and different approaches to characterize also the nature of landscape changes, to assess GIS-derived geomorphic indices and variables as described in Sect. 3.1 and to model the location and response of phenomena through interfaces [82, 84]. For instance, the use of digital elevation models derived by optical or radar remote sensing data within a GIS might provide the extraction of several morphological and hydrological parameters (e.g. altitude, slopes, orientation, basins and sub-basins, channel network, terraces, floodplain, curvature, aspect, flow direction, flow accumulation) as useful support for a change detection analysis. However, the consideration of an automated feature extraction approach remains controversial, because different variables could influence the reliability of this process [65]. Many authors described limits of automatic approaches to channel mapping including, for example, water level [66], vegetation cover, shadows and soil moisture [65, 69].

3.5 Application: Geomorphic Effects of the November 2013 Flood in the Posada Catchment, Italy

An example of application of optical remote sensing as a tool in channel change detection is illustrated in this section. The study referred to one single high-magnitude and low-frequency flood event that produced notable geomorphic response in two ungauged Mediterranean rivers in November 2013 (northeastern Sardinia, Italy) [72]. The study area is located in the Posada River basin which covers a total area of 685 km^2, and the analysis specifically focused on 22-km-long and 18-km-long semi-alluvial single-thread channels of the Posada and Mannu di Bitti Rivers, respectively.

The definition of channel morphological characteristics and features and the quantification of channel planimetric 2D changes (i.e. channel widening, floodplain planform change and island erosion/formation) through the use of high-resolution space-borne data within a geographic information system (GIS) are the main purposes of this remote sensing analysis. The choice of the study reaches is based on the fundamental assumption concerning the visibility of the channel and the feasibility of the assessment of geomorphic changes from above.

The evaluation of the initial morphological conditions deals with (1) the spatial units definition (i.e. from catchment scale to sub-reach scale) and (2) the quantitative measurements of the areal and linear geomorphic features [70] (i.e. channel width, channel slope, alluvial plain and island extension, valley confinement and channel sinuosity). The analysis on the pre flood high-resolution (1 m or higher) satellite images taken in April 2011 was carried out on data stored in the Basemap gallery available in ArcGIS Explorer Desktop by the World Imagery service (Fig. 7a). The planimetric 2D change detection was carried out by means of a multi-temporal analysis approach by visual skills-based interpretation procedure applied to a set of high-resolution pre- and post-flood satellite images and handled by the same operator. The post-flood image analysis was conducted on WorldView-2 pan-sharped 4 band imageries with a spatial resolution of 50 cm taken in April 2014 (Fig. 7b). The channel was digitized on pre- and post-satellite images, referring to the active-channel area, defined as the part of the river corridor relatively free of vegetation (unvegetated or sparsely vegetated bars) that conveys most of the water and sediment during high flow with an initial average width ranging from 9 to 141 m. Alluvial plain area encompassed present floodplain and low terraces, i.e. surfaces that can be some meters higher than the floodplain and can be infrequently flooded [80]. Islands were recognized and outlined as in-channel landforms that persists sufficiently long to establish permanent vegetation (i.e. shrub or woody vegetation).

Analysis findings revealed channel widening as the dominant geomorphic response observed. Channel width increased from 1.1 (i.e. from a pre-flood channel width of 141–156.5 m) to 6.2 (from a pre-flood channel width of 12.2–76 m) times the pre-flood width. In Fig. 7 an example of pre- and post-flood satellite imagery comparison at sub-reach scale (i.e. 800 m) is reported. In that sub-reach, the channel widened 5.5 times in comparison to the initial channel width (i.e. from 24 to 133 m).

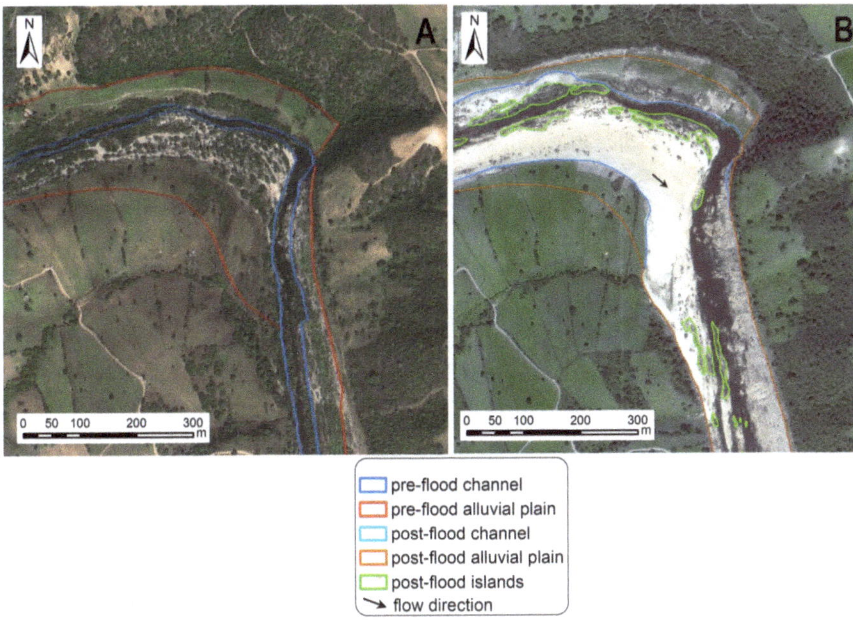

Fig. 7 Planimetric 2D changes (i.e. channel widening, floodplain planform change and reactivation and island formation) on the Posada River (northeastern Sardinia, Italy) at sub-reach scale: (**a**) 2011 pre-flood satellite image from Basemap gallery available in ArcGIS Explorer Desktop by the World Imagery service (spatial resolution = 1 m or higher); (**b**) 2014 post-flood WorldView-2 pan-sharped 4 band imagery (spatial resolution = 0.5 m)

Morphodynamic processes led to island formation and variation of large portions of floodplain planform. Specifically, in the sub-reach showed in Fig. 7, floodplain enlarged from an initial average width of 158 m to an average post-flood width of 196 m (i.e. 38 m of net change), while the formation of islands reached an areal extension of 10,346 km^2.

The spatial resolution of satellite imagery was suitable for the study purpose because pixel size was much smaller than river size and appropriate for fluvial features identification and assessment of channel change. The approach allowed continuous measurements and data supply over long distance in a large but limited access basin area. However, the analysis might be affected by errors, in particular related to imagery interpretation and digitization. Besides, different coordinate systems of pre (i.e. WGS 1984 Web Mercator Auxiliary Sphere)- and post-flood (i.e. UTM WGS 1984) satellite data originated small image shifting. The estimated error ranged from 0 to $\pm 16\%$ of change overestimation/underestimation, and it was obtained by calculating the difference between the maximum and the minimum shift values measured for each sub-reach. Errors were negligible considering that (1) images with high spatial resolution were used, (2) magnitude of changes was very high (i.e. the amount of changes in measured parameters largely exceeded errors) and (3) absolute values of pre- and post-flood geomorphic features allowed to overcome shifting errors and to obtain net change values. Furthermore, additional information was collected from digital elevation model (10 m spatial resolution), topographic maps at 1:10,000 scale, and field survey on representative reaches for minimizing errors, improving visual interpretation and obtaining a more robust analysis. Besides, the proximity of available data to the event allowed to minimize errors in considering possible natural geomorphic changes before or after the 2013 flood. Finally, the described remote sensing analysis allowed (1) a better understanding of channel dynamics and processes during high-magnitude flood event, (2) defining the potential controlling factors of channel widening (i.e. both morphological and hydraulic variables) and (3) developing predictive models describing magnitude and patterns of channel planimetric changes [72].

4 Discussion

As illustrated in the previous sections, several approaches may be adopted to detect geomorphic changes in response to floods, depending on the object of investigation. The feasibility of a certain approach depends on several aspects (e.g. systematic, random, georectification and digitization errors; the relationship between the size of the detecting feature and the spatial data resolution; magnitude of changes) which have different importance in relation to the focus of the geomorphic analysis. The most common approaches to change detection in response to floods, and their related data and feasibility, are summarized in Table 2. The best approach for channel width change analysis is a multi-temporal ad spatial analysis using high-resolution airborne or space-borne imagery, whose applicability might be mostly affected by the magnitude of changes and by the channel width related to image spatial resolution. For changes in channel width, georectification and digitization, errors can be considered of negligible or little importance in case of high-magnitude changes. Conversely, the two latter aspects are very relevant in the applicability of pre- and post-flood comparison approach for channel shift or migration and bank retreat detection. Besides, these analyses do not necessarily require field survey validations. Channel pattern and substrate-type change detection are usually conducted through qualitative multi-temporal and spatial analyses to support field survey; hence, the error assessment becomes less important, while the size of the detected object and the magnitude of the channel change remain necessary to reach a good matching between remotely sensed and field data. Analysis of exposed or submerged geomorphic features requires high-resolution multi- and hyperspectral imagery, respectively, to carry out a multi-temporal and spatial analysis. For exposed geomorphic feature detection, suitability of imagery comparison approaches might be highly affected by the minimum size of the detected features based on data resolution, magnitude of change and also georeferencing and digitization errors. A fair submerged feature detection, instead, depends mostly on spectral resolution

Table 2 Channel change detection analysis in response to floods: remotely sensed data, methods and main aspects to be taken into account for detecting changes

			Feasibility		
Change detection	Data	Method	Size of detected feature/data resolution	Change magnitude	Errors
Channel width	Aerial and satellite imagery	Multi-temporal and spatial analysis	+++	+++	+
Channel pattern	Aerial and satellite imagery	Multi-temporal and spatial analysis	+++	++	+
Channel shifting	Aerial and satellite imagery	Multi-temporal and spatial analysis	+++	++	+++
Bank retreat	Aerial and satellite imagery	Multi-temporal and spatial analysis	+++	+++	+++
Substrate type	Aerial and multi-spectral satellite imagery	Multi-temporal and spatial analysis; imagery comparison; field survey	+++	++	+
Grain size	Aerial, multi-spectral satellite and UAS imagery	Image analysis; field survey	+++	+	++
Submerged geomorphic feature	Airborne/space-borne hyperspectral imagery	Imagery classification comparison; field survey	++	++	+
Topography	Airborne/space-borne hyperspectral imagery	Bathymetric mapping	+	+++	+++
Erosion/deposition	LiDAR or photogrammetry derived DEMs/DTMs	DEM of Difference (DoD)	++	+++	+++
Land coverage	Aerial and multi-spectral satellite imagery	Multi-temporal and spatial analysis; imagery classification comparison	+	+	++

Feasibility of detecting geomorphic changes considering three aspects: the relationship (i.e. ratio) between size of the detected feature and spatial resolution of imagery, magnitude of geomorphic change, and errors due to georectification, digitization, etc.
The importance of each aspect is assessed as follows: + negligible or little importance, ++ moderate importance, +++ high or very high importance

(see Sect. 3.3). The grain size analysis necessitates the use of high-resolution airborne, space-borne or UAS imagery: in this case, availability of very high spatial resolution can be crucial for an effective analysis. Land cover change analysis is usually performed by classified imagery comparison, which does not really need georeferencing and digitization error minimization for a qualitative analysis. Nevertheless, this aspect might be subtle if accomplishing quantitative analysis. The definition of erosional or depositional process is mostly performed by the use of topographic data and by DoD approach. As described above, this kind of approach might be affected by several sources of error; however, determining the minimum level of detection (LoD) is fundamental, as well as the magnitude of change related to data spatial resolution.

5 Final Remarks

The present availability of data with high spatial resolution and short revisit time offers great opportunities and new challenges in the analysis of geomorphic response to floods: (1) the possibility of time-sequential satellite image comparison; (2) a broad areal coverage and a synoptic view of channel dynamics of individual flood event across different scales (i.e. from in-stream units to the entire catchment); (3) continuous measurements and data supply of channel response to flood of the whole river corridor over long distances, particularly in large fluvial systems and where accessibility is more difficult; (4) insights into the complexity of interactions driving changes in riverine landscape during flood events; and (5) deeper understanding of stream morphodynamics and potential to develop more accurate models of geomorphic response to floods to assess the magnitude of the flood event and areas of channel instability.

There are some key assumptions to be taken into account when optical remote sensing approach is used: (1) the temporal proximity of images to the flood event, for accurate analysis of the pre- and post-flood geomorphic characteristics; (2) the size of watershed and stream channels; (3) relationships between channel width, and other fluvial features, and the spatial resolution of the image; (4) magnitude of channel changes that must be well identifiable to capture in-stream variations; and (5) reduction of factors that may alter the quality of image interpretation or classification, such as seasonal stage, weather conditions and vegetation canopy. Furthermore, accuracy of optical remote sensing results is a still long debated aspect due to several sources of error.

Acknowledgements Fondazione Cariparo is gratefully acknowledged for funding M. Righini PhD fellowship. We thank the anonymous reviewers and the book editors for their comments and helpful suggestions.

References

1. Allen, G.H., Pavelsky, T.M.: Patterns of river width and surface area revealed by the satellite-derived north American River width data set. Geophys. Res. Lett. **42**, 395–402 (2015). https://doi.org/10.1002/2014GL062764
2. Baker, V.R.: Stream-channel response to floods, with examples from central Texas. Geol. Soc. Am. Bull. **88**(8), 1057–1071 (1977)
3. Bizzi, S., Lerner, D.N.: The use of stream power as an indicator of channel sensitivity to erosion and deposition processes. River Res. Applic. **31**, 16–27 (2013). https://doi.org/10.1002/rra.2717
4. Bledsoe, B.P., Watson, C.C.: Logistic analysis of channel pattern thresholds: Meandering, braiding, and incising. Geomorphology. **38**, 281–300 (2001)
5. Brasington, J., Rumsby, B.T., McVey, R.A.: Monitoring and modelling morphological change in a braided gravel-bed river using high-resolution GPS-based survey. Earth Surf. Process. Landf. **25**, 973–990 (2000)
6. Bryant, R.G., Gilvear, D.J.: Quantifying geomorphic and riparian land cover changes either side of a large flood event using airborne remote sensing: River Tay, Scotland. Geomorphology. **29**, 307–321 (1999)
7. Buraas, E.M., Renshaw, C.E., Magilligan, F.J., Dade, W.B.: Impact of reach geometry on stream channel sensitivity to extreme floods. Earth Surf. Process. Landf. **39**, 1778–1789 (2014)
8. Carbonneau, P.E., Piégay, H.: Introduction: the growing use of imagery in fundamental and applied river sciences. In: Carbonneau, P., Piégay, H. (eds.) Fluvial Remote Sensing for Science and Management, pp. 1–18. Wiley-Blackwell, Chichester, UK (2012)
9. Carbonneau, P.E., Piégay, H., Lejot, J., Dunford, R., Michel, K.: Hyperspatial imagery in riverine environments. In: Carbonneau, P., Piégay, H. (eds.) Fluvial Remote Sensing for Science and Management, pp. 163–191. Wiley-Blackwell, Chichester, UK (2012)
10. Cavalli, M., Trevisani, S., Comiti, F., Marchi, L.: Geomorphometric assessment of spatial sediment connectivity in small Alpine catchments. Geomorphology. **188**, 31–41 (2013). https://doi.org/10.1016/j.geomorph.2012.05.007
11. Chandler, J., Ashmore, P., Paola, C., Gooch, M., Varkaris, F.: Monitoring River-Channel change using terrestrial oblique digital imagery and automated digital photogrammetry. Ann. Assoc. Am. Geogr. **92**(4), 631–644 (2002). https://doi.org/10.1111/1467-8306.00308
12. Costa, J.E., O'Connor, J.E.: Geomorphically effective floods. In: Costa, J.E., Miller, A.J., Potter, K.W., Wilcock, P. (eds.) Natural and Anthropogenic Influences in Fluvial Geomorphology Monograph, vol. 89, pp. 45–56. American Geophysical Union, Washington, DC (1995)
13. Crosilla, F., Macorig, D., Scaioni, M., Sebastianutti, I., Visintini, D.: Lidar data filtering and classification by skewness and kurtosis iterative analysis of multiple point cloud data categories. Appl. Geomatics. **5**(3), 225–240 (2013)
14. Dethier, E., Magilligan, F.J., Renshaw, C.E., Nislow, K.H.: The role of chronic and episodic disturbances on channel–hillslope coupling: The persistence and legacy of extreme floods. Earth Surf. Process. Landf. **41**, 1437–1447 (2016). https://doi.org/10.1002/esp3958
15. Dewan, A.M., Islam, M.M., Kumamoto, T., Nishigaki, M.: Evaluating flood hazard for land-use planning in Greater Dhaka of Bangladesh using remote sensing and GIS techniques. Water Resour. Manag. **21**, 1601–1612 (2007). https://doi.org/10.1007/s11269–006–9116-1
16. Ferencevic, M.V., Ashmore, P.: Creating and evaluating digital elevation model-based stream-power map as a stream assessment tool. River Res. Appl. **28**, 1394–1416 (2012)
17. Fonstad, M.J.: Hyperspectral Imagery in Fluvial Environments. In: Carbonneau, P.E., Piégay, H. (eds.) Fluvial Remote Sensing for Science and Management. John Wiley & Sons, Ltd, Chichester, UK (2012). https://doi.org/10.1002/9781119940791.ch4
18. Fonstad, M.A., Marcus, W.A.: High resolution, basin extent observations and implications for understanding river form and process. Earth Surf. Process. Landf. **35**(6), 680–698 (2010)
19. Gilvear, D.J., Davids, C., Tyler, A.N.: The use of remotely sensed data on channel hydromorphology. River Tummel, Scotl. River Res. Applic. **20**, 1–17 (2004). https://doi.org/10.1002/rra792

20. Gilvear, D.J., Bryant, R.: Analysis of aerial photography and other remotely sensed data. In: Kondolf, G.M., Piégay, H. (eds.) Tools in fluvial geomorphology, pp. 133–168. Wiley, London (2003)
21. Gilvear, D., Winterbottom, S., Sichingabula, H.: Character of channel planform change and meander development: Luangwa River, Zambia. Earth Surf. Process. Landf. **25**, 421–436 (2000)
22. Ghoshal, S., James, L.A., Singer, M.B., Aalto, R.: Channel and floodplain change analysis over a 100-year period: Lower Yuba River, California. Remote Sens. **2**, 1797–1825 (2010). https://doi.org/10.3390/rs2071797
23. Grove, J.R., Croke, J., Thompson, C.: Quantifying different riverbank erosion processes during an extreme flood event. Earth Surf. Process. Landf. **38**, 1393–1406 (2013). https://doi.org/10.1002/esp3386
24. Gupta, A., Fox, H.: Effects of high magnitude floods on channel form: A case study in the Maryland Piedmont. Water Resour. Res. **10**, 499–509 (1974)
25. Gurnell, A.M.: Channel change of the river Dee meanders, 1946–1992, from the analysis of air photographs regulated rivers. Res. Manag. **13**, 13–26 (1997)
26. Heritage, G.L., Large, A.R.G., Moon, B.P., Jewitt, G.: Channel hydraulics and geomorphic effects of an extreme flood event on the Sabie River, South Africa. Catena. **58**, 151–181 (2004)
27. Hooke, J.M.: Cutoffs galore!: Occurrence and causes of multiple cutoffs on a meandering river. Geomorphology. **61**(3–4), 225–238 (2004)
28. Hooke, J.M.: Temporal variations in fluvial processes on an active meandering river over a 20-year period. Geomorphology. **100**, 3–13 (2008)
29. Hooke, J.M., Yorke, L.: Rates, distributions and mechanisms of change in meander morphology over decadal timescales, River Dane, UK. Earth Surf. Process. Landf. **35**, 1601–1614 (2010)
30. Hughes, M.L., McDowell, P.F., Marcus, W.A.: Accuracy assessment of georectified aerial photographs: Implications for measuring lateral channel movement in a GIS. Geomorphology. **74**, 1–16 (2006)
31. Jordan, D.C., Fonstad, M.A.: Two-Dimensional mapping of river bathymetry and power using aerial photography and GIS on the Brazos river, Texas. Geocarto Inter. **20**(3), 1–8 (2005)
32. Joyce, K.E., Belliss, S.E., Samsonov, S.V., McNeill, S.J., Glassey, P.J.: A review of the status of satellite remote sensing and image processing techniques for mapping natural hazards and disasters. Prog. Phys. Geogr. **33**(2), 1–25 (2009). https://doi.org/10.1177/0309133309339563
33. Krapesch, G., Hauer, C., Habersack, H.: Scale orientated analysis of river width changes due to extreme flood hazard. Nat. Hazards Earth Syst. Sci. **11**, 2137–2147 (2011)
34. Kumar, R., Kamal, V., Singh, R.K.: Geomorphic effects of 2011 floods on channel belt parameters of Rapti river: A remote sensing and GIS approach. Coron. J. Sci. Technol. **2**(II), 4–12 (2013)
35. Lane, S.N., Widdison, P.E., Thomas, R.E., Ashworth, P.J., Best, J.L., Lunt, I.A., Sambrook Smith, G.H., Simpson, C.J.: Quantification of braided river channel change using archival digital image analysis. Earth Surf. Process. Landf. **35**, 971–985 (2010)
36. Lane, S.N., Westaway, R.M., Hicks, D.M.: Estimation of erosion and deposition volumes in a large, gravel bed, braided river using synoptic remote sensing. Earth Surf. Process. Landf. **28**(3), 249–271 (2003)
37. Lane, S.N., Chandler, J.H., Porfiri, K.: Monitoring river channel and flume surfaces with digital photogrammetry. J. Hydraul. Eng. **127**, 871 (2001)
38. Lane, S., Richards, K., Chandler, J.: Developments in photogrammetry – the geomorphological potential. Prog. Phys. Geogr. **17**(3), 306–328 (1993)
39. Lea, D.M., Legleiter, C.J.: Refining measurements of lateral channel movement from image time series by quantifying spatial variations in registration error. Geomorphology. **258**, 11–20 (2016)
40. Legleiter, C.J., Roberts, D.A., Marcus, W.A., Fonstad, M.A.: Passive optical remote sensing of river channel morphology and in-stream habitat: Physical basis and feasibility. Remote Sens. Environ. **93**, 493–510 (2004)
41. Legleiter, C.J., Roberts, D.A., Lawrence, R.L.: Spectrally based remote sensing of river bathymetry. Earth Surf. Process. Landf. **34**, 1039–1059 (2009)

42. Legleiter, C.J., Fonstad, M.A.: An introduction to the physical basis for deriving river information by optical remote sensing. In: Carbonneau, P., Piégay, H. (eds.) Fluvial Remote Sensing for Science and Management, pp. 43–69. Wiley-Blackwell, Chichester, UK (2012)
43. Leopold, L.B., Maddock, T.J.: Hydraulic geometry of stream channels and some physiographic implications. U.S. Geol. Surv. Prof. Pap. **252**, 55 (1953)
44. Lichter, M., Klein, M.: The effect of river floods on the morphology of small river mouths in the southeastern Mediterranean Zeitschriftfür. Geomorphology. **55**(3), 317–340 (2011)
45. Lu, D., Mausel, P., Brondizio, E., Moran, E.: Change detection techniques. Int. J. Remote Sens. **25**(12), 2365–2407 (2004)
46. Magilligan, F.J., Buraas, E.M., Renshaw, C.E.: The efficacy of stream power and flow duration on geomorphic responses to catastrophic flooding. Geomorphology. **228**, 175–188 (2015)
47. Magilligan, F.J., Gomez, B., Mertes, L.A.K., Smith, L.C., Smith, N.D., Finnegan, D., Garvin, J.B.: Geomorphic effectiveness, sandur development, and the pattern of landscape response during jökulhlaups: Skeiðarársandur, southeastern Iceland. Geomorphology. **44**, 95–113 (2002)
48. Magilligan, F.J.: Thresholds and the spatial variability of flood power during extreme floods. Geomorphology. **5**, 373–390 (1992)
49. Marcus, W.A., Fonstad, M.A., Legleiter, C.J.: Management applications of optical remote sensing in the active river channel. In: Carbonneau, P., Piégay, H. (eds.) Fluvial Remote Sensing for Science and Management, pp. 19–41. Wiley-Blackwell, Chichester, UK (2012)
50. Marcus, W.A., Fonstad, M.A.: Optical remote mapping of rivers at sub-meter resolutions and watershed extents. Earth Surf. Process. Landf. **33**, 4–24 (2008). https://doi.org/10.1002/esp1637
51. Mertes, L.A.K.: Remote sensing of riverine landscapes. Freshw. Biol. **47**, 799–816 (2008)
52. Milan, D.J.: Geomorphic impact and system recovery following an extreme flood in an upland stream: Thinhope Burn, northern England, UK. Geomorphology. **138**, 319–328 (2012)
53. Miller, A.J.: Flood hydrology and geomorphic effectiveness in the central Appalachians. Earth Surf. Process. Landf. **15**, 119–113 (1990)
54. Milton, J.A., Gilvear, D.J., Hooper, I.D.: Investigating river channel changes using remote sensed data. In: Gurnell, A., Petts, G.E. (eds.) Changing River Channels, pp. 277–301. Wiley, Chichester (1995)
55. Mount, N.J., Louis, J.: Estimation and propagation of error in measurements of river channel movement from aerial imagery. Earth Surf. Process. Landf. **30**, 635–643 (2005)
56. Nardi, L., Rinaldi, M.: Spatio-temporal patterns of channel changes in response to a major flood event: The case of the Magra River (central-northern Italy). Earth Surf. Process. Landf. **40**, 326–339 (2014)
57. Ortega, J.A., Guillermina, G.H.: Geomorphological and sedimentological analysis of flash-flood deposits the case of the 1997 Rivillas flood (Spain). Geomorphology. **112**, 1–14 (2009)
58. Pan, Z., Glennie, C., Legleiter, C., Overstreet, B.: Estimation of water depths and turbidity from hyperspectral imagery using support vector regression. IEEE Geosci. Remote Sens. Lett. **12**(10), 2165–2169 (2015)
59. Peixoto, J.M.A., Nelson, B.W., Wittmann, F.: Spatial and temporal dynamics of river channel migration and vegetation in central Amazonian white-water floodplains by remote-sensing techniques. Remote Sens. Environ. **113**, 2258–2266 (2009)
60. Perignon, M.C., Tucker, G.E., Griffin, E.R., Friedman, J.M.: Effects of riparian vegetation on topographic change during a large flood event, Rio Puerco, New Mexico, USA. J. Geophys. Res. Earth Surf. **118**, 1193–1209 (2013). https://doi.org/10.1002/jgrf.20073
61. Piégay, H., Darby, S.E., Mosselman, E., Surian, N.: A review of techniques available for delimiting the erodible river corridor: A sustainable approach to managing bank erosion. River Res. Appl. **21**, 773–789 (2005)
62. Piégay, H., Kondolf, G.M., Minearc, J.T., Vaudora, L.: Trends in publications in fluvial geomorphology over two decades: A truly new era in the discipline owing to recent technological revolution? Geomorphology. **248**, 489–500 (2015)

63. Phillips, J.D.: Geomorphic impacts of flash flooding in a forested 828 headwater basin. J. Hydrol. **269**, 236–250 (2002)
64. Pradhan, B.: Flood susceptible mapping and risk area delineation using logistic regression, GIS and remote sensing. J. Spat. Hydrol. **9**(2), 1–18 (2010)
65. Priestnall, G., Aplin, P.: Spatial and temporal remote sensing requirements for river monitoring international. J. Remote Sens. **27**(11), 2111–2120 (2006)
66. Puech, C., Raclot, D.: Using geographical information systems and aerial photographs to determine water levels during floods. Hydrol. Process. **16**(8), 1593–1602 (2002)
67. Rango, A., Laliberte, A., Herrick, J.E., Winters, C., Havstad, K., Steele, C., Browning, D.: Unmanned aerial vehicle based remote sensing for rangeland assessment, monitoring, and management. J. Appl. Remote Sens. **3**, 1–15 (2009). https://doi.org/10.1117/1.3216822
68. Richardson, J.M., Fuller, I.C.: Quantification of Channel Planform Change on the Lower Rangitikei River, New Zealand, 1949–2007: Response to Management? Massey University School of People, Environment and Planning, Palmerston (2010)
69. Rinaldi, M., Simoncini, C., Piégay, H.: Scientific strategy design for promoting a sustainable sediment management: The case of the Magra River (Central – Northern Italy). River Res. Appl. **25**, 607–625 (2009). https://doi.org/10.1002/rra1243
70. Rinaldi, M., Surian, N., Comiti, F., Bussettini, M.: A method for the assessment and analysis of the hydromorphological condition of Italian streams: The Morphological Quality Index (MQI). Geomorphology. **180–181**, 96–108 (2013). https://doi.org/10.1016/jgeomorph201209009
71. Rinaldi, M., Amponsah, W., Benvenuti, M., Borga, M., Comiti, F., Lucìa, A., Marchi, L., Nardi, L., Righini, M., Surian, N.: An integrated approach for investigating geomorphic response to extreme events: Methodological framework and application to the October 2011 flood in the Magra River catchment, Italy. Earth Surf. Process. Landf. **41**, 835–846 (2016). https://doi.org/10.1002/esp.3902
72. Righini, M.: Geomorphic response to extreme floods in alluvial and semi-alluvial rivers, Unpublished PhD Thesis, University of Padua. (2017)
73. Sanyal, J., Lu, X.X.: Application of remote sensing in flood management with special reference to monsoon Asia: A review. Nat. Hazards. **33**, 283–301 (2004)
74. Schumm, S.A., Lichty, R.W.: Channel widening and floodplain construction along Cimarron River in southwestern Kansas. U.S. Geol. Surv. Prof. Pap. **352D**, 71–88 (1963)
75. Singh, A.: Review article. Digital change detection techniques using remotely-sensed data. Int. J. Remote Sens. **10**(6), 989–1003 (1989). https://doi.org/10.1080/01431168908903939
76. Sloan, J., Miller, J.R., Lancaster, N.: Response and recovery of the Eel River, California, and its tributaries to floods in 1955, 1964 and 1997. Geomorphology. **36**, 129–154 (2001)
77. Smith, L.C.: Satellite remote sensing of river inundation area, stage, and discharge: A review. Hydrol. Process. **11**, 1427–1439 (1997)
78. Stweart, J.H., La Marche, V.C. Jr: Erosion and deposition produced by the flood of December 1964 on Coffee Creek, Trinity County, California. U.S. Geol Survey Prof Paper 422K (1967)
79. Surian, N., Rinaldi, M., Pellegrini, L., Audisio, C., Maraga, F., Teruggi, L., Turitto, O., Ziliani, L.: Channel adjustments in northern and central Italy over the last 200 years. In: James, L.A., Rathburn, S.L., Whittecar, G.R. (eds.) Management and Restoration of Fluvial Systems with Broad Historical Changes and Human Impacts Special Paper, 451, pp. 83–95. Geological Society of America, Boulder (2009)
80. Surian, N., Righini, M., Lucìa, A., Nardi, L., Amponsah, M., Benvenuti, M., Borga, M., Cavalli, M., Comiti, F., Marchi, L., Rinaldi, M., Viero, A.: Channel response to extreme floods: Insights on controlling factors from six mountain rivers in northern Apennines, Italy. Geomorphology. **272**, 78–91 (2016). https://doi.org/10.1016/j.geomorph.2016.02.002
81. Tamminga, A.D., Eaton, B.C., Hugenholtz, C.H.: UAS-based remote sensing of fluvial change following an extreme flood event. Earth Surf. Proc. Landf. **40**, 1464–1476 (2015). https://doi.org/10.1002/esp3728
82. Tholey, N., Clandillon, S., De Fraipont, P.: The contribution of spaceborne SAR and optical data in monitoring flood events: Examples in northern and southern France. Hydrol. Process. **11**, 1409–1413 (1997)

83. Thompson, C., Croke, J.: Geomorphic effects, flood power, and channel competence of a catastrophic flood in confined and unconfined reaches of the upper Lockyer valley, southeast Queensland, Australia. Geomorphology. **197**, 156–169 (2013)
84. Walsh, S.J., Butler, D.R., Malanson, G.P.: An overview of scale, pattern, process relationships in geomorphology: A remote sensing and GIS perspective. Geomorphology. **21**, 183–205 (1998)
85. Wohl, E.: Time and the rivers flowing: Fluvial geomorphology since 1960. Geomorphology. **216**, 263–282 (2014)
86. Wolman, M.G., Gerson, R.: Relative scales of time 854 and effectiveness of climate in watershed geomorphology. Earth Surf. Proc. Landf. **3**, 189–208 (1978)
87. Wolman, M.G., Miller, J.P.: Magnitude and frequency of forces in geomorphic processes. J. Geol. **68**, 54–74 (1960)
88. Woodget, A., Carbonneau, P., Visser, F., Maddock, I.: Quantifying submerged fluvial topography using hyperspatial resolution UAS imagery and structure from motion photogrammetry. Earth Surf. Process. Landf. **40**(1), 47–64 (2015)
89. Yang, X., Damen, M.C.J., Van Zuidam, R.A.: Satellite remote sensing and GIS for the analysis of channel migration changes in the active Yellow River Delta, China. JAG l. **1**(2), 146–157 (1999)

The Use of DEM-Based Approaches to Derive a Priori Information on Flood-Prone Areas

Salvatore Manfreda, Caterina Samela, and Tara J. Troy

Abstract Knowing the location and the extent of areas exposed to floods is the most basic information needed for planning flood management strategies. Unfortunately, a complete identification of these areas is still lacking in many countries. Recent studies have highlighted that a significant amount of information regarding the inundation process is already contained in the structure and morphology of a river basin. Therefore, several geomorphic approaches have been proposed for the delineation of areas exposed to flood inundation using DEMs. Such DEM-based approaches represent a useful tool, characterized by low cost and simple data requirements, for a preliminary identification of the flood-prone areas or to extend flood hazard mapping over large areas. Moreover, geomorphic information may be used as external constraint in remote-sensing algorithms for the identification of inundated areas during or after a flood event.

Keywords Flood hazard maps • DEM-based approaches • Binary classifiers

1 Introduction

The delineation of areas exposed to flood hazard raises complex problems regarding the definition of flood events and the parameterization of hydraulic models (e.g. [27, 61]). The increasing damages caused by recent floods and weather-related events forced river basin authorities and environmental agencies to invest significant efforts in flood mapping. This need is motivated by the fact that more and more people live in flood-prone areas [7] but also by the increasing frequency of extreme floods (e.g. [19, 47]).

S. Manfreda (✉) • C. Samela
Department of European and Mediterranean Cultures: Architecture, Environment, Cultural Heritage (DiCEM), Università degli Studi della Basilicata, Potenza, Italy
e-mail: salvatore.manfreda@unibas.it

T.J. Troy
Lehigh University, Bethlehem, PA, USA

A. Refice et al. (eds.), *Flood Monitoring through Remote Sensing*, Springer Remote Sensing/Photogrammetry,
https://doi.org/10.1007/978-3-319-63959-8_3

Making and maintaining an accurate flood hazard mapping is neither simple nor inexpensive. In fact, these maps represent the result of painstaking topographic surveys and heavy numerical hydraulic simulation of flood wave propagation (e.g. [26]). In this context, topographic data are the most important factor in determining flow direction, water surface elevations and extent of flooding.

On one hand, dynamics of the flooding process clearly indicate the existence of an intrinsic link between floodplain extent and topography. On the other hand, landscape evolution results from the complex interplay of many variables, such as climate, hydrology, geology, sediment load, vegetation, valley dimensions and human activity. In fact, the morphology of a river basin keeps track of all past flood events that accelerate the processes of erosion, transport and deposition and over long periods of time are able to shape and form geomorphic features [44]. Thus, one might expect that meaningful information on flooding can be extracted from features of landscape morphology (e.g. [13, 22, 25, 45, 50, 63]).

River basin morphology can be characterized by automatic procedures using digital elevation models (DEMs). The availability of a variety of DEMs (e.g. ASTER GDEM, SRTM, GMTED2010, EU-DEM) has given a strong impulse to the development of the so-called DEM-based models. These models aim at supplying procedures for the delineation of areas exposed to flood inundation by using the basin's morphological characteristics derived from DEMs.

2 DEM-Based Approaches for the Delineation of Flood-Prone Areas

The original idea that the hydrological responses of a catchment could be indexed on the basis of topography dates back to Kirkby [34]. He proposed the use of the TOPMODEL topographic index, defined as $\ln(a/\tan\beta)$, which combines the local upslope contributing area per unit contour length (a) and slope angle of ground surface ($\tan\beta$). It represents an index of convergence of the subsurface flow and it is well correlated with spatial patterns of soil moisture. In fact, areas characterized by high specific contributing area or minor characteristic slope have higher values of topographic index and therefore higher probability to be saturated. This index is commonly used to quantify the topographic control in numerous hydrological models [3, 5, 39, 48] and is also known to be a good indicator of flood-prone areas [9, 32, 40, 52].

Its ability in reproducing the flood-prone areas was improved by Manfreda et al. [40] by changing the relative weight of the drained area with respect to the local slope with an exponent $n < 1$. This exponent was introduced in order to provide a measure of the relative value assumed by the hydraulic radius $\left(\sim A_d^n\right)$ in a given point. Manfreda et al. [40] compared this 'modified topographic index' (TI_m) with flood inundation maps obtained from hydraulic simulations and observed that the portion of a basin exposed to flood inundation can be detected applying a threshold value, τ, on the map of TI_m.

This paper led to a number of applications where geomorphic indices, such as the TI and TI_m, are adopted for the delineation of flood-prone areas (e.g. [6, 10, 11, 32, 35, 51, 52]). In some of these studies, the procedure developed by Manfreda et al. [40] and the relative GRASS GIS tool [12] have been simply applied. In other studies, the procedure has been modified exploring new calibration strategies. For instance, Jalayer et al. [32] used a Bayesian parameter estimation in order to characterize the uncertainties in delineating the flood-prone areas related to the evaluation of the threshold. In the remaining cases, the TI or TI_m indices have been used in combination with other features.

Apart from the use of the modified topographic index, the potential use of geomorphologic features for the delineation of flood-prone areas has been explored in an increasing number of papers. Starting from simple information about basin features, these methods have been significantly improved with time, reaching good accuracy.

In this context, Mehlhorn et al. [46] suggested the use of non-linear regression of easily available geomorphologic catchment attributes, such as distance to river and upstream catchment area. They proposed a tool [60] that has been implemented within a flood hazard mapping system for insurance purposes.

Williams et al. [62] developed a DEM-based floodplain delineation method based on the intersection between the topographic surface and a water level of reference. They assumed a constant water level for the entire network, which is compared to the elevation of locations contributing to each stream cell. Basin cells with elevations lower than the water depth are part of the inundated area. This method has some drawbacks, such as an overestimation of water depth, especially at upstream locations, and a lack of consistency with the basin hydrologic conditions.

Nardi et al. [49] overcame these limitations by proposing a hydrogeomorphic delineation procedure able to calculate from terrain analyses a variable flood stage for each stream cell with the associated extent of surrounding inundated area, which is related to the geomorphic properties of the stream network and the hydrologic characteristics of a flood event.

Degiorgis et al. [11] proposed a procedure for the identification of the areas subject to flooding hazard starting from remotely sensed elevation data and existing flood hazard maps. They used linear binary classifiers to investigate the relationships between morphology and flooding hazard at the catchment scale. In their work, the classifiers are based on five selected quantitative morphological features derived from DEMs. Several performance measures were used to identify the best classifier and to validate the procedure, such as the area under the receiver operating characteristic (ROC) curve, the true positive rate, the false positive rate, and their combination. According to their application over the Tanaro basin (Italy), the best-performing feature among those selected is the difference in elevation (H) between the location under examination and the downstream river node to which the site is hydrologically connected. This feature showed a high degree of accuracy, correctly identifying 91% of the flood-exposed areas with an overestimation error of 16%, with higher errors in the flat portion of the basin.

Manfreda et al. [41] carried out a comparative analysis of three geomorphic approaches proposed by Nardi et al. [49], Manfreda et al. [40] and Degiorgis et al. [11], respectively, on two sub-catchments of the Tiber river in Central Italy. The work emphasized the main characteristics of the investigated methodologies, revealing that methods based on morphological indices provided a good description of the flooded areas. It also confirmed the ability of a single feature in characterizing the flood hazard, such as the difference in elevation (H).

This study provided a preliminary investigation on the role played by some morphological features on flood exposure and represents the starting point of the studies carried out by Manfreda et al. [42, 43] and Samela et al. [58, 59], which took further steps forward regarding the number of descriptors investigated, the size and relevance of the study area and the validation of the procedure for large-scale applications, with the objective to identify the most meaningful geomorphologic attributes for flood inundation processes. The final purpose is to use such descriptors to map the flood exposure in data-poor environments and/or over large spatial scales with a cost-effective method that adopts freely available worldwide information (e.g. DEMs).

3 Methods and Material

3.1 Digital Elevation Models

The development of several hydrological tools received a significant impulse by the increasing availability of detailed DEMs. The gridded DEM is the most commonly used in environmental modelling since it can be easily stored and manipulated [21, 31, 53]. Typically, DEMs are derived by photogrammetric data, satellite image data, laser scanning data, cartographic data and ground surveys [20].

A variety of DEMs have been recently produced by new remote-sensing techniques providing a better representation of the Earth's surface. As a consequence of the growing availability of global DEMs, they have been massively used as main data source for research purposes in the field of geomorphology and hydraulics [23, 65]. As a result, DEMs with a resolution of 30–90 m are available from various sources over the entire globe. Some examples are the Advanced Spaceborne Thermal Emission and Reflection Radiometer-Global Digital Elevation Model (ASTER GDEM) [54], the Shuttle Radar Topography Mission (SRTM) [17] and the Global Multi-Resolution Terrain Elevation Data 2010 (GMTED2010), followed by newly released improved versions (e.g. NASA SRTM v3, CGIAR-CSI SRTM v4.1, DLR/ASI SRTM X-SAR DEM, ASTER GDEM v2 and EU-DEM). The DEMs cover most of the populated regions of the world, with a minimum spatial resolution of 1 arc s for SRTM and ASTER, and 7.5 arc s for the GMTED2010, and several researchers tested and evaluated their accuracy in different parts of the world (e.g. [1, 24, 55, 64]).

For hydrological applications, the DEMs of HydroSHEDS (Hydrological Data and Maps Based on Shuttle Elevation Derivatives at Multiple Scales, [28]) developed by the Conservation Science Program of the World Wildlife Fund can be used. This dataset offers a suite of georeferenced hydrographic information for regional- and global-scale applications with the great advantage of being homogeneous in quality and freely available. This product is based primarily on elevation data obtained during NASA's Shuttle Radar Topography Mission (SRTM) generated using the synthetic aperture radar (SAR) technique [17], with ancillary data sources including the SRTM Water Body Data [29], the river networks of the Digital Chart of the World (DCW) [14] and ArcWorld [15] and the Global Lakes and Wetlands Database [36]. SRTM provided a DEM available almost in near-global coverage, mapping all land surfaces between 60° north latitude and 56° south latitude. Unfortunately, in its original release, the SRTM data contain regions with missing data point (voids or anomalies) and a large number of sinks or depressions that can cause significant problems (see [16]). Instead, HydroSHEDS provides continuous DEMs, so-called void-filled elevation model (DEM VOID), where voids have been filled, and all not natural sinks and main elevation inconsistencies have been identified and removed. Besides sinks, the original DEMs show a series of other anomalies, like the influence of vegetation cover [56] that can cause significant errors in the derived river courses and flow directions. Therefore, a hydrologically conditioned elevation model (DEM CON) is also available in order to establish continuous flow for hydrological applications. This model is obtained after a sequence of hydrologic conditioning procedures, forcing the DEM to produce correct river network topology and to produce a DEM-derived drainage network coherent with the actual one. The processing steps of generating HydroSHEDS are detailed in the dataset's technical documentation [37]. Preliminary quality assessments indicate that the accuracy of HydroSHEDS significantly improved that of existing global watershed and river maps. Information about the data quality and shortcomings of HydroSHEDS data can be found in Lehner et al. [38].

Note that since the conditioning process of the DEM-CON significantly alters the original elevation data, its use should be limited only to the drainage network identification procedures, while the quantitative measures of morphological characteristics (e.g. local slope, curvature or elevation difference between points) should be derived from the original DEM-VOID data [42].

Figure 1 provides a visual example of SRTM DEM of the Bradano river (located in southern Italy) that will be used throughout the paper as reference case study.

3.2 Measures of Performance

The potential of DEM-based geomorphic procedures can be tested using standard flood maps obtained with more traditional models (e.g. hydraulic simulations, field

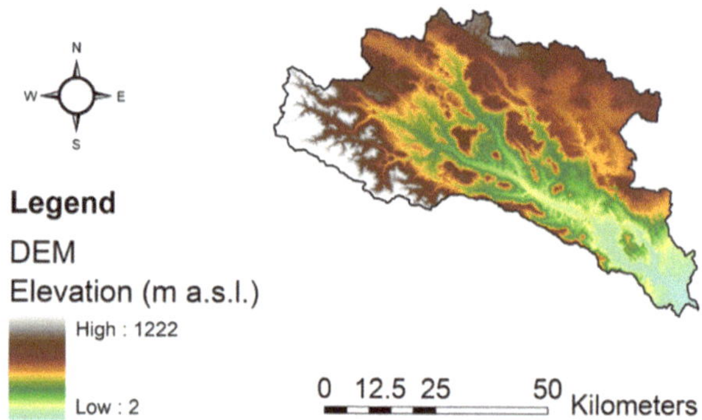

Fig. 1 SRTM DEM of the Bradano river basin (Basilicata, Southern Italy)

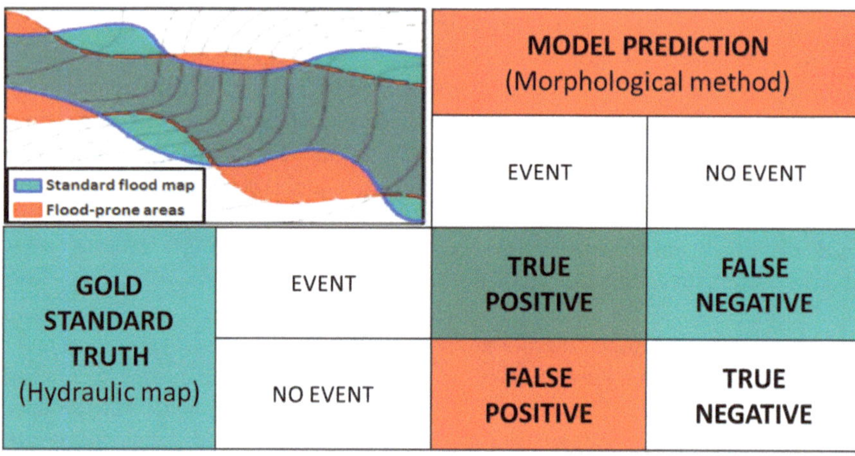

Fig. 2 Overview of the four possible outcomes of a multiple statistic comparison between a test prediction and a gold standard truth

measurements or remotely sensed maps). For a comparison between a gold standard truth and a model prediction, we can use four metrics, as shown in the contingency table of Fig. 2:

- True positives (TP): basin locations that the model correctly identifies as flooded and that are inundated in the standard flood map as well.
- True negatives (TN): basin locations that the model prediction correctly identifies as "not flooded" and that are not flooded in the standard flood hazard map.
- False positives (FP): basin locations that the model prediction incorrectly identifies as flooded, but are not flooded in the flood hazard map. This is an error of overestimation, a false alarm.
- False negatives (FN): basin locations that the model prediction incorrectly identifies as "not flooded", but are flooded in the flood hazard map. This is an error of underestimation.

Given the above definitions, the standard metrics generally used are:

- *True positive rate*: R_{TP} = TP/(TP + FN) ⟶ Sensitivity.
  It defines how many correct positive results occur among all positive samples available during the test.
- *False negative rate*: R_{FN} = FN/(TP + FN) = 1 − R_{TP} ⟶ Underestimation (error type II).
  It defines how many incorrect negative results occur among all positive samples available during the test.
- *True negative rate*: R_{TN} = TN/(TN + FP) ⟶Specificity.
  It defines how many correct negative results occur among all negative samples available during the test.
- *False positive rate*: R_{FP} = FP/(TN + FP)= 1 − R_{TN} ⟶ Overestimation (error type I).
  It defines how many incorrect positive results occur among all negative samples available during the test.

4 Linear Binary Classifiers Based on Geomorphological Indices

Linear binary classifiers seek to separate the datasets in two classes by using a "linear boundary" which is a hyperplane in the space; in two dimensions, a hyperplane corresponds to a simple line identified by the value of a threshold. Once the threshold is found, we can use it to predict the class a new point belongs to by simply checking on which side of the boundary it falls. Such a methodology can be used to identify areas subject to flooding hazard using several morphological features.

Starting from morphological descriptors presumed to be good flood-related candidates, several classifiers have been built to delineate flood-prone areas based solely on the information contained in DEMs. The values of these descriptors usually vary widely and differ from one descriptor to the other. In order to apply thresholds that cover the whole range of values, it is convenient to standardize them through normalization. Specifically, the values are scaled and translated so that all data assume a value between −1 and 1. Consequently, normalized thresholds τ are applied and iteratively changed over a portion of the basin that plays the role of calibration area. Every threshold represents the line that divides the values in two classes, distinguishing flood-prone and non-flood-prone areas. Therefore, for every threshold a potential flood-prone area binary map is obtained and compared to a standard flood map assumed as "gold standard truth". These standard maps have been previously converted into binary maps, where the value 0 represents the areas not flooded (area of minimal flood hazard) and the value 1 represents the flooded areas. Classifier calibration is carried out minimizing the sum of the false positive rate and the false negative rate (R_{FP} + R_{FN}) in order to derive the optimal value of the normalized threshold.

As performance measures, necessary to evaluate the reliability of each classifier, we adopted the statistic metrics introduced in the previous section, which have been used to draw the ROC curves. They are defined as the set of pairs of true positive rate (plotted on the y-axis) and false positive rate (plotted on the x-axis) and are obtained by varying the threshold of the classifier [18]. The diagonal line $y = x$ represents the strategy of randomly guessing a class. Any classifier that appears in the higher left triangle performs better than random selection. Furthermore, in order to compare different binary classifiers, the calculation of the area under the ROC curve (AUC) has been used. The value of the AUC ranges from 0.5 (completely random classifier) to 1.0 (perfectly discriminating classifier).

From an operational point of view, the relationships between the selected morphological features and the flood map are first calibrated at the sample scale and then applied to extend the hazard information at the basin scale.

Several applications and tests have been performed, considering different hydrologic, climatic and topographic contexts, different scale of analysis and different input data in terms of DEM resolution, standard flood maps adopted (derived by 1D or 2D models) and dominant topography of the training area.

5 Detection of Flood-Prone Areas

With the aim to deepen our understanding on the potential of geomorphic descriptor and to define new strategies for preliminary flood hazard detection, the mentioned classifiers have been tested in several river basins located in different parts of the world (Europe, the USA and Africa). Among the numerous DEM-derived morphological features investigated in previous studies, two have proven to be representative of flood exposure:

- The distance of each cell from the nearest stream, D [m], defined as the length of the path hydrologically connecting the location under exam to the nearest element of the drainage network.
- The Geomorphic Flood Index (GFI), $\ln(h_r/H)$: this index compares in each point of the basin a variable water depth h_r with the elevation difference H. In this case, $h_r \approx bA_r^n$ is computed as a function of the contributing area A_r in the nearest section of the drainage network hydrologically connected to the point under exam ('r' stands for 'river') (see Fig. 3). By taking into account an estimate of the water level in the nearest element of the drainage network h_r, we are considering the nearest river as the hazard source. In this way, this index tries to describe the level of safety of a given site with respect to the probable water level of the river due to the relative position of that site and the magnitude of the drained basin.

The GFI is specifically developed by Manfreda et al. [43] and Samela et al. [58, 59] with the specific aim of identifying a morphological feature able to account for the main features affecting flood diffusion. It represents a dimensionless ratio

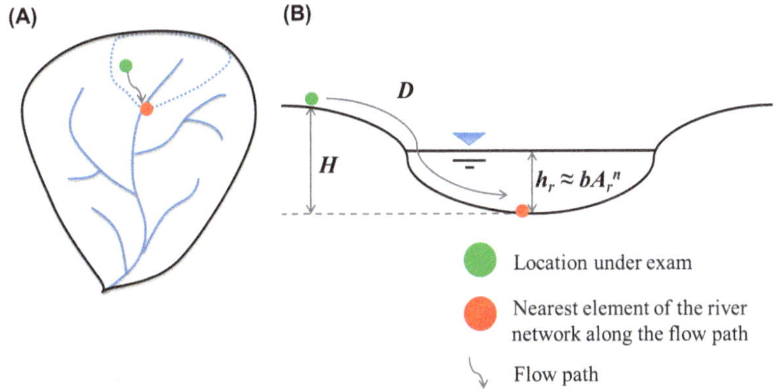

Fig. 3 Description of the composite index GFI. Representation of the parameters H and h_r in plan (**a**) and in the cross-section (**b**)

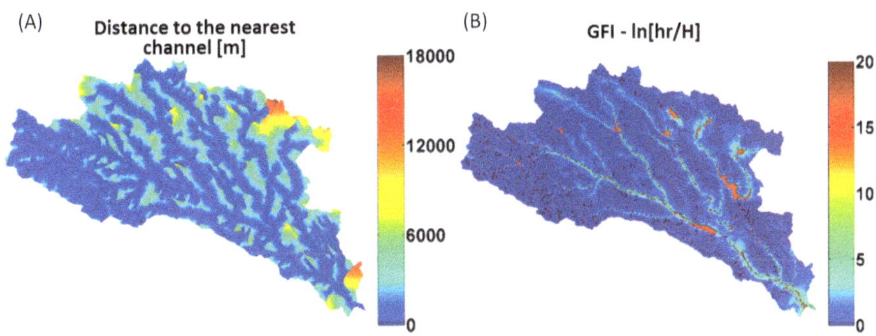

Fig. 4 DEM-derived morphological descriptors of the Bradano river basin (Basilicata, Italy): (**a**) distance to the nearest channel, D; (**b**) the Geomorphic Flood Index, GFI

between the potential water depth and the relative elevation of a location. Therefore, it increases in areas potentially inundated or more likely to be flooded.

Figure 4 provides a visual description of the two indices introduced above for the Bradano river basin (Basilicata, Southern Italy). These two indices have an opposite behaviour with the first (D) decreasing when the considered point is closer to the river network (source of risk), while the second increases its value. This means that the binary classifier will consider as flooded all points below a given threshold for the classifier D or above for the classifier GFI.

An example of application of a linear binary classifier for the delineation of the flood-prone areas is depicted in Fig. 5 for the Bradano river. This graphs provide a visual comparison of a flood map obtained with a hydraulic model and the one obtained with this DEM-based approach. The DEM-based flood map reproduces fairly well the hydraulic flood map with the clear advantage of extending the flood mapping in the portion of the river basin not studied with hydraulic models.

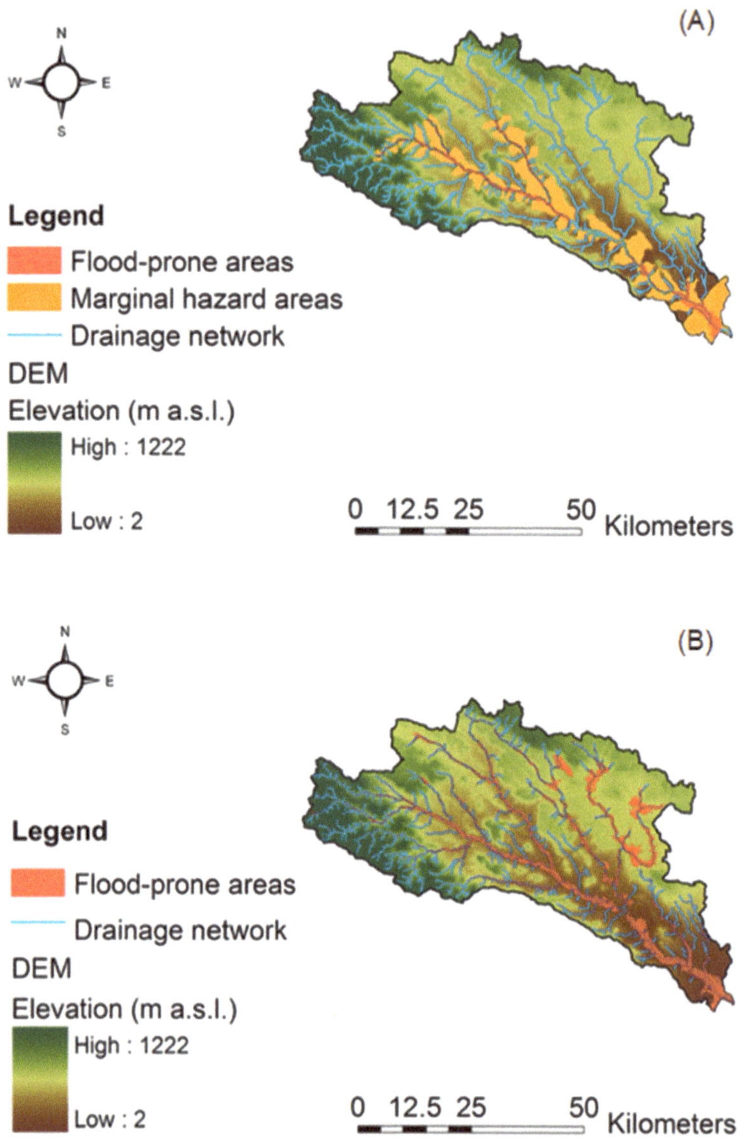

Fig. 5 (**a**) Flood hazard map derived by using a one-dimensional hydraulic modelling over the Bradano river obtained with a 30-year return period. (**b**) Flood susceptibility map obtained by using the linear binary classifier based on the Geomorphic Flood Index, $\ln(h_r/H)$, on the entire Bradano river basin

This kind of analysis has been carried out extensively over a wide range of basins characterized by significant differences in terms of climate, morphology and size. Table 1 summarizes the outcomes of the linear binary classification applied to 21 river basins located in different parts of the world and obtained by using the mentioned DEM-derived morphological descriptors.

It is necessary to remark that each application has been carried out using different reference flood maps for calibration. In particular, the Flood Insurance Rate Maps (FIRMs) developed by the Federal Emergency Management Agency (FEMA) have been adopted to train the classifiers in the US basins. Typically, these FIRMs are derived by hydraulic analyses, but in some cases, approximate methods of analysis are also used. For our purpose, flood inundation maps having a 500-year return period have been used, excluding coastal areas that may result flooded due to storm surge.

For the basins located in Italy, flood hazard maps produced by river basin authorities have been adopted. Usually, these maps are the result of intense studies regarding the main river and its most important tributaries, while many minor tributaries are yet unstudied. Different return periods (T) have been chosen, such as $T = 200$ years for the upper Tiber river basin; $T = 500$ years for the basins of Basento, Sinni, Cavone, Noce and Agri; and $T = 30$ years for the Bradano basin.

Finally, for the ungauged subbasin of the Bulbula river (Ethiopia), a flood hazard map has been first derived by performing a hydrologic/hydraulic analysis with a 300-year return period and then used to calibrate the classifiers.

These reference flood maps have been computed over several different basins, by different authors and also with different methodologies. Nevertheless, the main interest is to explore the reliability of DEM-based procedures in describing flood-prone areas rather than derive flood maps with a specific return period or with a specific reliability.

The performance measures reported in Table 1 have been computed within the calibration areas defined by the standard flood hazard map mentioned. According to the results of our studies, the GFI provides better performances, in terms of minimization of the total error ($R_{FP} + R_{FN}$) in 16 of 21 study cases. The accuracy of the maps derived with the GFI has a mean R_{TP} of about 82.1% (with a range of values between 74.8% and 98.9%) with a R_{FN} of about 17.9% (ranging from 14.3% to 39.5%). By contrast, the maps derived with D have a mean R_{TP} of 78.4% (ranging from 45.8% to 97.8%) with a R_{FN} of 21.6% (ranging from 10.4% to 41.5%). Therefore, GFI seems to be more consistent and less sensitive to the change of scale of application, DEM resolution, reference hydraulic map used for calibration and different topography of the training area. In a few cases, the feature D provides slightly better performances, and this is generally observed in flat areas. Given this result, one may decide which classifier to use according to the mean slope of the considered river basin.

The ensemble of the performances observed seems to endorse the classifier based on the GFI as the most satisfactory among those analysed, and the most suitable to detect the flood-prone areas in ungauged basins and for large-scale applications, providing good accuracies with low data requirements.

It is also interesting to remark the limited range of variability of the calibrated thresholds within a specific region. This may be extremely useful to extend the application of the methodology in ungauged basins.

Table 1 Performances of the parameter D and the index GFI in all the basins investigated across the world

		Flow distance, D					Geomorphic Flood Index, $\ln(h_r/H)$				
River basin	Drainage area [km^2]	τ	R_{TP}	R_{FP}	$R_{FP} + R_{FN}$	AUC	τ	R_{TP}	R_{FP}	$R_{FP} + R_{FN}$	AUC
Ohio (USA)[a]	529,000	−0.985	63.9%	20.9%	57.0%	0.777	−0.515	81.6%	20.1%	38.5%	0.891
Delaware (USA)[a]	35,253	−0.930	54.6%	19.7%	65.1%	0.716	−0.510	76.7%	24.7%	48.0%	0.845
Upper Guyandotte (USA)[a]	2,481	−0.945	89.5%	27.8%	38.2%	0.864	−0.625	91.6%	18.7%	27.2%	0.917
Lower Guyandotte (USA)[a]	1,937	−0.935	79.7%	22.0%	42.3%	0.849	−0.590	86.7%	23.1%	36.5%	0.897
Tug (USA)[a]	4,078	−0.945	86.1%	24.9%	38.8%	0.862	−0.635	89.9%	21.1%	31.2%	0.910
Upper Levisa (USA)[a]	3,111	−0.945	86.8%	28.5%	41.6%	0.846	−0.610	89.1%	17.4%	28.3%	0.915
Big Sandy (USA)[a]	1,095	−0.915	79.8%	24.8%	45.1%	0.844	−0.550	85.1%	17.0%	31.9%	0.913
Tippecanoe (USA)[a]	5,158	−0.880	86.1%	28.7%	42.6%	0.860	−0.570	77.8%	33.7%	55.9%	0.762
Middle Wabash-Little Vermilion (USA)[a]	6,031	−0.905	79.0%	23.5%	44.6%	0.846	−0.515	74.8%	15.5%	40.8%	0.853
Vermilion (USA)[a]	3,836	−0.905	82.9%	18.2%	35.3%	0.889	−0.480	68.6%	16.6%	48.0%	0.826
Sugar (USA)[a]	2,198	−0.915	76.2%	21.0%	44.8%	0.840	−0.525	75.0%	21.6%	46.5%	0.832
Eel (USA)[a]	3,216	−0.895	73.9%	29.0%	55.1%	0.802	−0.585	78.6%	21.0%	42.4%	0.846
Upper Tiber (Italy)[b]	5,000	−0.977	77.5%	22.4%	44.9%	0.848	−0.379	85.9%	22.2%	36.3%	0.898
Bradano basin – 1D hydraulic model (Italy)[c]	2,765	−0.928	78.4%	17.8%	39.5%	0.882	−0.422	93.9%	14.3%	20.4%	0.950
Bradano outlet – 2D hydraulic model (Italy)[c]	2,765	−0.842	45.8%	24.7%	78.9%	0.625	−0.423	81.1%	26.7%	45.6%	0.791
Agri (Italy)	1,686	−0.779	76.3%	37.4%	61.1%	0.765	−0.396	78.8%	29.8%	51.0%	0.795
Basento (Italy)	1,537	−0.864	96.8%	41.5%	44.7%	0.800	−0.344	84.3%	19.5%	35.2%	0.885
Sinni (Italy)	1,292	−0.811	74.9%	37.9%	63.0%	0.748	−0.424	78.6%	39.5%	60.9%	0.748
Cavone (Italy)	648	−0.777	81.4%	33.7%	52.3%	0.801	−0.378	69.1%	30.0%	60.9%	0.737
Noce (Italy)	378	−0.854	79.7%	40.4%	60.8%	0.763	−0.45	77.3%	29.2%	51.9%	0.776
Subbasin of the Bulbula (Ethiopia)[d]	85	−0.968	97.8%	10.4%	12.6%	0.971	−0.512	98.9%	13.9%	15.0%	0.969

[a]Samela et al. [59]
[b]Manfreda et al. [42]
[c]Manfreda et al. [43]
[d]Samela et al. [58]

6 Discussion

DEM-based approaches may provide a cost-efficient strategy for the delineation of flood-prone areas. The tools described herein are mainly based on the use of a linear binary classification based on the recently proposed Geomorphic Flood Index (GFI) and the parameter D, which exhibited high classification accuracy and reliability in several test sites located in Europe, the USA and Africa [43, 57–59].

DEM-based methods only rely on the morphological characteristics of a river basin, assuming that these features are mainly controlled by basin hydrology. The threshold parameters display a limited variability within a specific region. Nevertheless, some changes are observed that may be due to local climate and geology. Therefore, their application requires a reference flood map for the calibration of the threshold parameter. Even considering such limitation, it is possible to apply the DEM-based method exploiting flood maps obtained with simplified hydrological/hydraulic modelling referring to a portion of the basin of interest (e.g. [58]).

Therefore, the most promising application for DEM-based approaches is oriented to extend the inundation classification providing a large-scale representation of flood exposure. This is possible in several cases across the world, since inundation mapping is available in most of the river basins, but the greatest limitation is due to the fact that it is time consuming to generate a complete mapping including all streams of a river network. Therefore, the method may be considered an excellent strategy to fill the gap of the actual flood mapping. An example of gap-filling application is given in Fig. 6, where the method is applied on the Ohio river basin (USA) providing the opportunity to extend the flood mapping in areas not studied by the Federal Emergency Management Agency (FEMA).

With the specific aim to make available to all users the proposed procedure that includes a number of operations that require good geomorphic and GIS competences, we recently implemented a Matlab code for GFI calculation [30] and an open-source QGIS plug-in named Geomorphic Flood Area tool (GFA tool) that adopts the linear binary classification based on the GFI. It allows to extend the flood classification outside a calibration area to delineate the flood-prone areas across the selected river basin. This plug-in may be extremely useful for flood mapping all over the world providing a powerful tool for all developing countries where classical flood studies are not affordable.

In this context, an aspect that deserves attention is the minimum extent of the training area for the application of this kind of methods. Samela et al. [59] studied the variability of the threshold parameter with the size of the training area and observed that the training area should contain more than 2% of the total basin area in order to obtain good performances.

We recall that the flood maps used to calibrate the classifiers have a specific return period; hence, the thresholds found and the related performances are sensitive to the considered return period of the training flood map. Theoretically, with the increase of the return period, the calibrated threshold should increase for the parameter D and decrease for the GFI.

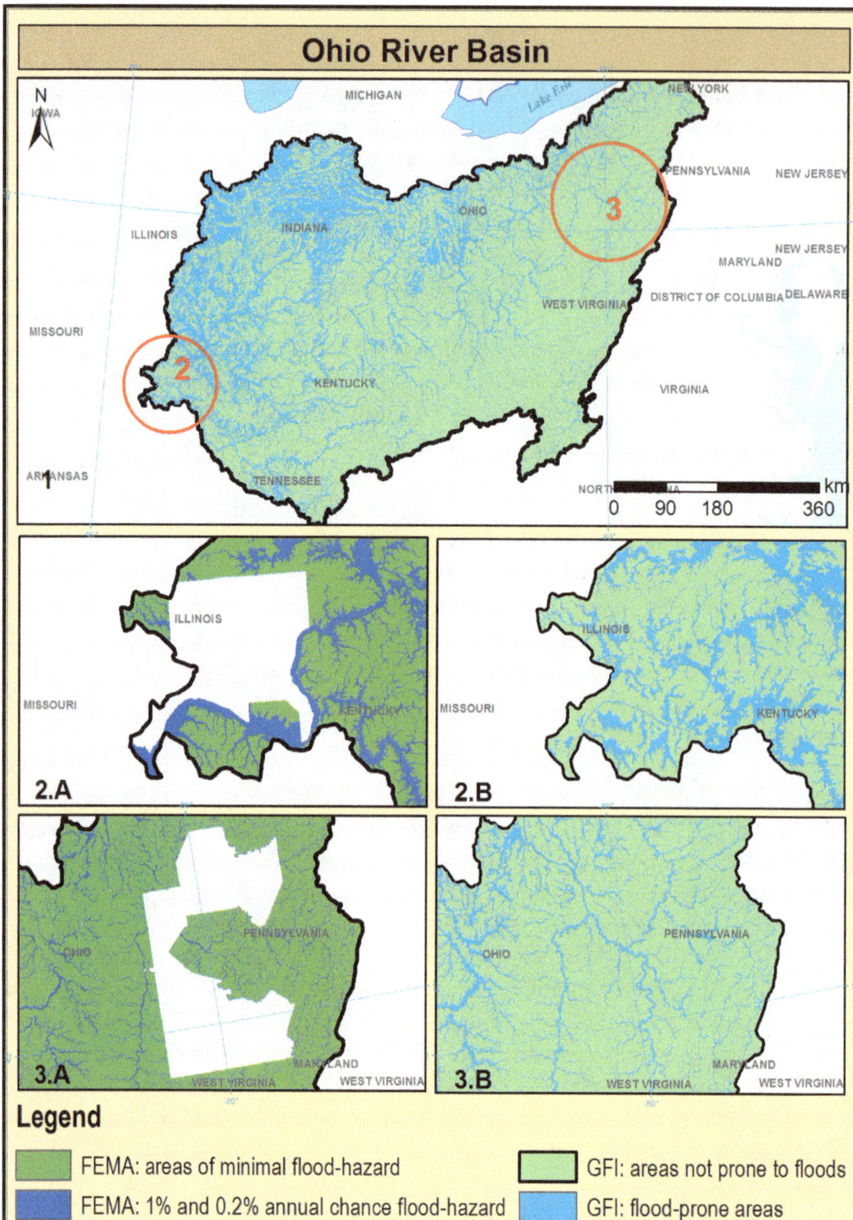

Fig. 6 Flood-prone areas (depicted in *light blue*) of the Ohio river basin according to the linear binary classifier based on the GFI (panel *1*). Comparison between the FEMA's flood map (panel *2.A* and *3.A*) and the flood map obtained using the proposed DEM-based approach (panel *2.B* and *3.B*)

Moreover, the proposed features provide indications that may be used in other contexts, such as in remote-sensing applications, where some a priori information at large scale (like those provided by the *GFI*) can be used to identify areas most likely to be flooded and help the design of new modelling schemes (e.g. [8]).

Finally, it is necessary to remark that the DEM-based method proposed in this work cannot substitute for a comprehensive study on the flood hazard, since geomorphologic analysis does not take into account the presence of flood defence structures, such as levees and dams. In addition, processes like levee failure or local scour may alter flood propagation in a way that is difficult to predict. Such scenarios deserve specific attention and are out of reach for such simple method. Notwithstanding a slight loss in accuracy, the proposed approach offers an economic and fast alternative for floodplain mapping. Such flood maps may be useful for large-scale flood preparedness and emergency response operations [33].

7 Conclusions

Floods account for about one-third of the economic losses and over half of the deaths associated with natural disasters worldwide [2]. The exposure of human activities is increasing with the expansion of cities, and, furthermore, there is a motivated suspect that the frequency of great floods is increasing with time [47]. This imposes in several countries the definition of procedures able to provide extended description of the flood-prone areas in the next few years.

In ungauged or poorly gauged sites, the lack of data does not permit comprehensive hydrologic/hydraulic studies about the flooding hazard exposure and deprives local authorities of reliable flood inundation maps. As a consequence, many communities are currently living in scarcely monitored basins exposed to flooding without any adequate assessments of areas potentially affected by the hazard.

In this scenario, geomorphic approaches represent a practical and economical procedure to identify preliminarily the extent of flood-prone areas. This can be achieved through an in-depth analysis of the basin morphology based on information freely available worldwide, such as DEMs.

The most appropriate morphologic signatures for the delineation of the flood-prone areas can be considered the GFI and the *D*. These classifiers have been tested on several river basins highlighting that the index GFI consistently exhibited higher accuracies among the others in each test and is the most suitable to detect the flood-prone areas in data-poor environments and for large-scale applications; it provided good accuracy with low requirements in terms of data and computational costs. The parameter *D* generally performs better in flat areas.

Geomorphic methods may also be used to discriminate between inundation maps derived for different return periods. Its application can lead to an extended flood map obtained using for calibration a portion of the entire river basin. Samela et al. [59] highlighted that a portion of the basin of approximately 2% may be sufficient to this aim.

This kind of approach may be extremely beneficial in the definition of new procedures for the identification of flooded areas from remote sensing techniques, where the topographic information may be used as an external constraint in the adopted algorithms that generally rely only on the local slope or the distance from the channel (e.g. [4]). Therefore, the integration of these geomorphic methods and the images obtained by remote sensing techniques can be used in order provide an accurate interpretation of the situation in the aftermath of an extreme hydrological event. A preliminary example of such an approach is given in D'Addabbo et al. [8], where this geomorphic method is used as ancillary data in flood mapping algorithms based on remote-sensed data to monitor flooded areas.

References

1. Athmania, D., Achour, H.: External validation of the ASTER GDEM2, GMTED2010 and CGIAR-CSI-SRTM v4. 1 free access digital elevation models (DEMs) in Tunisia and Algeria. Remote Sens. **6**(5), 4600–4620 (2014). doi.org/10.3390/rs6054600
2. Benz, G.: Flood disasters: lessons from the past – worries for the future. Proc. Inst. Civ. Eng. Water Marit. Energy. **142**(1), 3–8 (2000)
3. Beven, K.J., Kirkby, M.J.: A physically based, variable contributing area model of basin hydrology. Hydrol. Sci. Bull. **24**, 43–69 (1979). doi.org/10.1080/02626667909491834
4. Brivio, P.A., Colombo, R., Maggi, M., Tomasoni, R.: Integration of remote sensing data and GIS for accurate mapping of flooded areas. Int. J. Remote Sens. **23**(3), 429–441 (2002). doi.org/10.1080/01431160010014729
5. Burt, T.P., Butcher, D.P.: Development of topographic indices for use in semi-distributed hillslope runoff models. In: Baltenau, D., Slaymaker, O. (eds.) Geomorphology and Land Management, vol. 58, pp. e1–e19. Z., Geomorphol. Suppl. Band (1986)
6. Cooper, R.T.: Open data flood mapping of Chao Phraya River basin and Bangkok metropolitan region. Br. J. Environ. Clim. Chang. **4**(2), 186 (2014)
7. Cutter, S.L., Emrich, C.: Are natural hazards and disaster losses in the US increasing? EOS. Trans. AGU. **86**(41), 381–389 (2005). doi.org/10.1029/2005EO410001
8. D'Addabbo, A., Refice, A., Pasquariello, G., Lovergine, F., Capolongo, D., Manfreda, S.: A Bayesian network for flood detection combining SAR imagery and ancillary data. IEEE Trans. Geosci. Remote Sens. **54**(6), 3612–3625 (2016). doi.org/10.1109/TGRS.2016.2520487
9. De Risi, R., Jalayer, F., De Paola, F., Iervolino, I., Giugni, M., Topa, M.E., Mbuya, E., Kyessi, A., Manfredi, G., Gasparini, P.: Flood risk assessment for informal settlements. Nat. Hazards. **69**(1), 1003–1032 (2013). doi.org/10.1007/s11069-013-0749-0
10. De Risi, R., Jalayer, F., De Paola, F.: Meso-scale hazard zoning of potentially flood prone areas. J. Hydrol. **527**, 316–325 (2015). doi.org/10.1016/j.jhydrol.2015.04.070
11. Degiorgis, M., Gnecco, G., Gorni, S., Roth, G., Sanguineti, M., Taramasso, A.C.: Classifiers for the detection of flood-prone areas using remote sensed elevation data. J. Hydrol. **470**, 302–315 (2012). doi.org/10.1016/j.jhydrol.2012.09.006
12. Di Leo, M., Manfreda, S., Fiorentino, M.: An automated procedure for the detection of flood prone areas: r.hazard.flood. Geomatics Workbooks n. 10 – "FOSS4G-it: Trento 2011", pp. 83–89. (2011). doi.org/10.1474/GGA.2005-01.0-19.0019
13. Dodov, B.A., Foufoula-Georgiou, E.: Floodplain morphometry extraction from a high-resolution digital elevation model: a simple algorithm for regional analysis studies. Geosci. Remote. Sens. Lett. IEEE. **3**(3), 410–413 (2006). doi.org/10.1109/LGRS.2006.874161
14. Environmental Systems Research Institute (ESRI): Digital chart of the world. 1:1M. Redlands (1993)
15. ESRI – Environmental Systems Research Institute: ArcWorld 1:3 Mio. Continental Coverage. ESRI, Redlands, CA. Data available on CD (1992)

16. Falorni, G., Teles, V., Vivoni, E.R., Bras, R.L., Amaratunga, K.S.: Analysis and characterization of the vertical accuracy of digital elevation models from the Shuttle Radar Topography Mission. J. Geophys. Res. Earth Surf. (2003–2012). **110**(F2), F02005. doi.org/10.1029/2003JF000113
17. Farr, T.G., Rosen, P.A., Caro, E., Crippen, R., Duren, R., Hensley, S., Kobrick, M., Paller, M., Rodriguez, E., Roth, L., Seal, D., Shaffer, S., Shimada, J., Umland, J., Werner, M., Oskin, M., Burbank, D., Alsdorf, D.: The shuttle radar topography mission. Rev. Geophys. **45**(2), RG2004 (2007). doi.org/10.1029/2005RG000183
18. Fawcett, T.: An introduction to ROC analysis. Pattern Recogn. Lett. **27**(8), 861–874 (2006). doi.org/10.1016/j.patrec.2005.10.010
19. Fischer, E.M., Knutti, R.: Observed heavy precipitation increase confirms theory and early models. Nat. Clim. Chang. **6**(11), 986–991 (2016). doi.org/10.1038/nclimate3110
20. Forkuo, E.K.: Digital terrain modelling in a GIS environment. Int. Arch. Photogramm. Remote. Sens. Spat. Inf. Sci. **37**(B2), 1023–1029 (2008)
21. Forkuo, E.K.: Digital elevation modelling using ASTER stereo imagery. J. Environ. Sci. Eng. **52**(2), 81–92 (2010)
22. Gallant, J.C., Dowling, T.I.: A multiresolution index of valley bottom flatness for mapping depositional areas. Water Resour. Res. **39**(12), 1347 (2003). doi.org/10.1029/2002WR001426
23. Gong, L., Halldin, S., Xu, C.Y.: Global-scale river routing – an efficient time-delay algorithm based on hydroSHEDS high-resolution hydrography. Hydrol. Process. **25**(7), 1114–1128 (2011). doi.org/10.1002/hyp.7795
24. Hirt, C., Filmer, M.S., Featherstone, W.E.: Comparison and validation of the recent freely available ASTER-GDEM ver1, SRTM ver4. 1 and GEODATA DEM-9S ver3 digital elevation models over Australia. Aust. J. Earth Sci. **57**(3), 337–347 (2010). doi.org/10.1080/08120091003677553
25. Hjerdt, K.N., McDonnell, J.J., Seibert, J., Rodhe, A.: A new topographic index to quantify downslope controls on local drainage. Water Resour. Res. **40**(5), W05602 (2004). doi.org/10.1029/2004WR003130
26. Horritt, M.S.: Calibration and validation of a 2-dimensional finite element flood flow model using satellite radar imagery. Water Resour. Res. **36**(11), 3279–3291 (2000). doi.org/10.1029/2000WR900206
27. Horritt, M.S., Bates, P.D.: Evaluation of 1D and 2D numerical models for predicting river flood inundation. J. Hydrol. **268**(1–4), 87–99 (2002). doi.org/10.1016/S0022-1694(02)00121-X
28. http://hydrosheds.cr.usgs.gov/index.php
29. http://edc.usgs.gov/products/elevation/swbdguide.doc
30. http://it.mathworks.com/matlabcentral/fileexchange/60815-geomorphic-flood-index
31. Hunter, G.J., Goodchild, M.F.: Modeling the uncertainty of slope and aspect estimates derived from spatial databases. Geogr. Anal. **29**(1), 35–49 (1997). doi.org/10.1111/j.1538-4632.1997.tb00944.x
32. Jalayer, F., De Risi, R., De Paola, F., Giugni, M., Manfredi, G., Gasparini, P., Topa, M.E., Yonas, N., Yeshitela, K., Nebebe, A., Cavan, G., Lindley, S., Printz, A., Renner, F.: Probabilistic GIS-based method for delineation of urban flooding risk hotspots. Nat. Hazards. **73**(2), 975–1001 (2014). doi.org/10.1007/s11069-014-1119-2
33. Kappes, M.S., Gruber, K., Frigerio, S., Bell, R., Keiler, M., Glade, T.: The Multi-RISK platform: the technical concept and application of a regional-scale multihazard exposure analysis tool. Geomorphology. **151**, 139–155 (2012)
34. Kirkby, M.J.: Hydrograph Modelling Strategies, pp. 69–90. Department of Geography. University of Leeds (1975)
35. Knitter, D., Bergner, M., Horejs, B., Schütt, B., Meyer, M.: Concepts of Centrality and Models of Exchange in Prehistoric Western Anatolia. eTopoi. Journal for Ancient Studies. ISO 690 (2012)
36. Lehner, B., Döll, P.: Development and validation of a global database of lakes, reservoirs and wetlands. J. Hydrol. **296**(1–4), 1–22 (2004). doi.org/10.1016/j.jhydrol.2004.03.028

37. Lehner, B., Verdin, K., Jarvis, A.: HydroSHEDS Technical Documentation, Version 1.0, pp. 1–27. World Wildlife Fund, Washington, DC (2006). (Available at http://hydrosheds.cr.usgs.gov)
38. Lehner, B., Verdin, K., Jarvis, A.: New global hydrography derived from spaceborne elevation data. Eos. Trans. AGU. **89**(10), (2008). doi.org/10.1029/2008EO100001
39. Manfreda, S., Fiorentino, M., Iacobellis, V.: DREAM: distributed model for runoff et antecedent soil moisture simulation. Adv. Geosci. **2**, 31–39 (2005)
40. Manfreda, S., Di Leo, M., Sole, A.: Detection of flood prone areas using digital elevation models. J. Hydrol. Eng. **16**(10), 781–790 (2011). doi.org/10.1061/(ASCE)HE.1943-5584.0000367
41. Manfreda, S., Nardi, F., Samela, C., Grimaldi, S., Taramasso, A.C., Roth, G., Sole, A.: Investigation on the use of geomorphic approaches for the delineation of flood prone areas. J. Hydrol. **517**, 863–876 (2014a). doi.org/10.1016/j.jhydrol.2014.06.009
42. Manfreda, S., Samela, C., Sole, A., Fiorentino, M: Flood-prone areas assessment using linear binary classifiers based on morphological indices. In: Vulnerability, Uncertainty, and Risk Quantification, Mitigation, and Management, pp. 2002–2011. ASCE (2014b). doi.org/10.1061/9780784413609.201
43. Manfreda, S., Samela, C., Gioia, A., Consoli, G.G., Iacobellis, V., Giuzio, L., Cantisani, A., Sole, A.: Flood-prone areas assessment using linear binary classifiers based on flood maps obtained from 1D and 2D hydraulic models. Nat. Hazards 1–20 (2015). doi.org/10.1007/s11069-015-1869-5
44. de Matauco, A.I.G., Ojeda, A.O., Bea, E.D.: Influence of catchment processes on fluvial morphology and river habitats. Limnetica. **30**(2), 169–182 (2011)
45. McGlynn, B.L., Seibert, J.: Distributed assessment of contributing area and riparian buffering along stream networks. Water Resour. Res. **39**(4), WR001521 (2003). doi.org/10.1029/2002WR001521
46. Mehlhorn, J., Feyen, L., Banovsky, I., Menzinger, I.: FRAT1. 0-an example of applying the geomorphologic regression approach for detailed single location flood risk assessment. In: Geophysical Research Abstracts, vol. 7, p. 07419. (2005)
47. Milly, P.C.D., Wetherald, R.T., Dunne, K.A., Delworth, T.L.: Increasing risk of great floods in a changing climate. Nature. **415**(6871), 514–517 (2002). doi.org/10.1038/415514a
48. Moore, I.D., Grayson, R.B., Ladson, A.R.: Digital terrain modelling: a review of hydrological, geomorphological, and biological applications. Hydrol. Process. **5**, 3–30 (1991). doi.org/10.1002/hyp.3360050103
49. Nardi, F., Vivoni, E.R., Grimaldi, S.: Investigating a floodplain scaling relation using a hydrogeomorphic delineation method. Water Res. Res. **42**(9), W09409 (2006). doi.org/10.1029/2005WR004155
50. Noman, N.S., Nelson, E.J., Zundel, A.K.: Review of automated floodplain delineation from digital terrain models. J. Water Resour. Plan. Manag. **127**(6), 394–402 (2001). doi.org/10.1061/(ASCE)0733-9496(2001)127:6(394)
51. Papaioannou, G., Vasiliades, L., Loukas, A.: Multi-criteria analysis framework for potential flood prone areas mapping. J Water Resour. Manag. Springer, Netherlands (2015). doi.org/10.1007/s11269-014-0817-6
52. Pourali, S.H., Arrowsmith, C., Chrisman, N., Matkan, A.A., Mitchell, D.: Topography wetness index application in flood-risk-based land use planning. Appl. Spat. Anal. Policy. **9**(1), 39–54 (2016)
53. Raaflaub, L.D., Collins, M.J.: The effect of error in gridded digital elevation models on the estimation of topographic parameters. Environ. Model. Softw. **21**(5), 710–732 (2006). doi.org/10.1016/j.envsoft.2005.02.003
54. Reuter, H.I., Nelson, A., Strobl, P., Mehl, W., Jarvis, A.: A first assessment of Aster GDEM tiles for absolute accuracy, relative accuracy and terrain parameters. In: Geoscience and Remote Sensing Symposium. 2009 IEEE International. IGARSS 2009, vol. 5, pp. V-240. IEEE (2009)
55. Rexer, M., Hirt, C.: Comparison of free high-resolution digital elevation data sets (ASTER GDEM2, SRTM v2. 1/v4. 1) and validation against accurate heights from the Australian National Gravity Database. Aust. J. Earth Sci. **61**(2), 213–226 (2014). doi.org/10.1080/08120099.2014.884983

56. Rodriguez, E., Morris, C.S., Belz, J.E.: A global assessment of the SRTM performance. Photogramm. Eng. Remote. Sens. **72**(3), 249–260 (2006). doi.org/10.14358/PERS.72.3.249
57. Samela, C.: 100-year flood susceptibility maps for the continental U.S. derived with a geomorphic method. University of Basilicata. Dataset (2016). 10.4121/uuid: 4dc7ef9f-80d0-4b78-8161-ca35f215d547
58. Samela, C., Manfreda, S., Paola, F. D., Giugni, M., Sole, A., Fiorentino, M.: DEM-based approaches for the delineation of flood-prone areas in an ungauged Basin in Africa. J. Hydrol. Eng. 06015010 (2016). https://doi.org/10.1061/(ASCE)HE.1943-5584.0001272
59. Samela, C., Troy, T.J., Manfreda, S.: Flood hazard mapping over large regions. Adv. Water Resour. **102**, 13–28 (2017). doi.org/10.1016/j.advwatres.2017.01.007
60. SwissRe.: Available at: http://www.swissre.com/clients/insurers/propertycasualty/SwissReGlobalFloodZonesenablingbetterbusinessdecisions.html (2012)
61. Werner, M., Blazkova, S., Petr, J.: Spatially distributed observations in constraining inundation modelling uncertainties. Hydrol. Process. **19**, 3081–3096 (2005). doi.org/10.1002/hyp.5833
62. Williams, W.A., Jensen, M.E., Winne, J.C., Redmond, R.L.: An automated technique for delineating and characterizing valley-bottom settings. Environ. Monit. Assess. **64**(1), 105–114 (2000). doi.org/10.1023/A:1006471427421
63. Wolman, M.G.: Evaluating alternative techniques floodplain mapping. Water Resour. Res. **7**(6), 1383–1392 (1971). doi.org/10.1029/WR007i006p01383
64. Yamazaki, D., Baugh, C.A., Bates, P.D., Kanae, S., Alsdorf, D.E., Oki, T.: Adjustment of a spaceborne DEM for use in floodplain hydrodynamic modeling. J. Hydrol. **436**, 81–91 (2012). doi.org/10.1002/2013WR014664
65. Yan, K., Tarpanelli, A., Balint, G., Moramarco, T., Baldassarre, G.D.: Exploring the potential of SRTM topography and radar altimetry to support flood propagation modeling: Danube case study. J. Hydrol. Eng. **20**(2), 04014048 (2014)

River Flood Forecasting System: An Interdisciplinary Approach

Viacheslav Zelentsov, Ilya Pimanov, Semen Potryasaev, Boris Sokolov, Sergey Cherkas, Andrey Alabyan, Vitaly Belikov, and Inna Krylenko

Abstract The chapter presents a holistic system that implements an advanced river flood modeling and forecasting approach. This approach extends traditional methods based on separate satellite monitoring or river physical processes modeling, by integration of different technologies such as satellite and in situ data processing, input data clustering and filtering, digital mapping of river valleys relief, data crowdsourcing, hydrodynamic modeling, inundation visualization, and also duly warning of stakeholders.

The software of the suggested system was implemented on the base of open source code and service-oriented architecture (SOA). This allows the use of different program modules for data processing and modeling, integrated into a unified software suite. Forecast results are available as web services. Additionally, a special GIS platform has been developed to visualize the results of forecasting. It does not require the users to have any special skills or knowledge, and all the complexity relating to data processing and modeling is hidden from the users.

The results of case studies have shown that the suggested interdisciplinary approach provides highly accurate forecasting due to operational ingestion and integrated processing of the remote sensing and ground-based water flow data in real time. In these case studies, forecasting of flood areas and depths was performed on a time interval of 12–48 h, allowing performing the necessary steps to alert and evacuate the population.

In general, the developed software can be considered as a toolkit to create holistic monitoring and flood forecasting systems with an integrated use of diverse satellite and in situ data.

The original version of this chapter was revised. An erratum to this chapter can be found at https://doi.org/10.1007/978-3-319-63959-8_9

V. Zelentsov (✉) • I. Pimanov • S. Potryasaev • B. Sokolov • S. Cherkas
Saint Petersburg Institute of Informatics and Automation of the Russian Academy of Sciences, St. Petersburg, Russia
e-mail: v.a.zelentsov@gmail.com

A. Alabyan • V. Belikov • I. Krylenko
Lomonosov Moscow State University, Moscow, Russia

A. Refice et al. (eds.), *Flood Monitoring through Remote Sensing*, Springer Remote Sensing/Photogrammetry, https://doi.org/10.1007/978-3-319-63959-8_4

Keywords River floods • Interdisciplinary approach • Multimodel simulation • Operational forecasting • Remote sensing data • Intelligent interface • Service oriented architecture

1 Brief Overview of Existing Systems and Services for Flood Forecasting

Large cities and small settlements all over the world have experienced more than once the destructive effects of floods. The number of floods is forecast to continue in the foreseeable future, and there is an emerging trend of growth of the total number of large-scale natural hydrological disasters [1–3]. In the current situation, the forecasting of floods caused by spring floods and freshets of various origins is one of the main components of safety providing for the settlements located in river valleys, and it is an important scientific problem [4].

By now a number of systems and information services have been developed and successfully used in emergency situations.

The United States Geological Survey (USGS) on its "WaterWatch" website [5] daily collects and updates data from US hydrological information collected at stream gauges, about current consumptions and levels of water (for 30 and more years), as well as separately about drought and flood events. For each site, results of observations statistical processing – monthly, annual, and multiyear average values of the observed parameters – can be obtained.

The products of Danish Hydrological Institute (DHI) for intensive monitoring of flood [6, 7] are special forecast and monitoring services for marine and freshwater objects (processes), where hydrological forecasts (locally upgradeable and adaptable in every case under the requirements and needs of customer-water users) are implemented.

The European Flood Awareness System [8] is a European project based on the LISFLOOD hydrological model [9], describing the runoff from a catchment area and the flow in the channels system. The model has been specifically designed for large watersheds and implemented by using GIS technologies.

Deltares Hydrological Institute, sited in Delft, the Netherlands, has a system similar to the MIKE Flood Watch product. It is framed as a project, by the Delft Flood Early Warning Systems [10], implementing real-time monitoring and management of water resources, which can be adapted to the needs of each area of interest. FEWS has been tested for flood forecasting and warning, presenting to its users a wide range of tools for data processing and management, as well as an interface that allows to apply different models.

Other existing models are, e.g., DHI Modeling complexes MIKE 11 [11] and MIKE 21 [12], multilayer two-dimensional model Delft 3D Deltares (Delft, the Netherlands) [13], model HEC-RAS [14], and FLO-2D [15], the body of the American Corp of Military Engineers, and the French model TELEMAC [16].

The Weather Research and Forecast [17] model, developed by the US National Center for Atmospheric Research, is used by ROSHYDROMET [18] and many other organizations to solve a wide range of weather forecasting issues in real time, as well as a dynamic core of computer modeling systems designed to study air transportation for hydrological forecasts. The WRF model contains flexible non-hydrostatic equations that take into account Earth surface heterogeneity: it implements a technology for calculations of enclosed areas with one-way or two-way influence; the model uses different coordinate systems depending on the type of surface.

The LANCE service [19] provides the following data sets: MODIS corrected reflectance, True Color, and Bands 7-2-1 and MODIS land surface reflectance, Bands 1-2-1, and AIRS precipitation (day/night). MODIS data is used to produce daily forecasts, connected with weather-related systems of coastal areas: MODIS Water Vapor (day/night), Cloud Top Temperature (day/night), and Cloud Top Pressure (day/night).

A further example of information product is the GIO EMS Mapping [20] for proper response to floods which requires immediate action (rush mode). For authorized users, three standard categories of cartographic information products are proposed: reference maps (used as basic information), delineation maps (providing an estimate of the scale of emergencies), and grading maps (providing spatial coverage of emergencies and assessment of the damage occurred). All maps can be provided in the form of overview maps or detailed maps. In addition, GIO EMS Mapping provides for potential customers (users):

- Full-color maps printing (ISO A1) or their equivalents in GeoPDF file format with high (300 dpi), medium (200 dpi), or low (100 dpi) resolution, as well as metadata files
- Geospatial full-color maps ISO A1 or its equivalent in GeoTIFF and georeferenced JPEG global geotagging for file formats with high (300 dpi), medium (200 dpi), or low (100 dpi) resolution, as well as metadata files (which may be the same as printed maps)
- Vector files for all the objects obtained in the analysis and interpretation of ESRI shapefiles, with the projection file (.prj), Google Earth KML (or KMZ), as well as the metadata file
- Radar satellites imageries (TerraSAR-X, COSMO-SkyMed, and RADARSAT)
- Optoelectronic satellites imageries (WorldView, GeoEye, Ikonos, Quickbird, SPOT, Formosat, RapidEye, and the Pleiades)
- Additional thematic information about infrastructure facilities, population, etc.
- DEM/DTM for satellite imagery orthorectification and risks analysis

In its turn, DigitalGlobe provides information products: available levels of Advanced Elevation Series [21] – map, accurate, and precision data DEM or DTM in 32-bit GeoTiff file format, as well as metadata ISO 19115: 2003.

The Global Land Cover Facility [22] allows to carry out analyses of the landscape at all levels from local to global. DEM data are available for the entire globe (GLSDEM-2008) and derived from radar imagery (SRTM-2000).

The following custom services (information products) are available from ISRO Remote Sensing Centre (India) [23]:

- Mapping and monitoring of floods in real time
- Assessment of flood damage
- Mapping of risks zones caused by possible rivers channels changes

The Centre is going to create flood early warning maps and to act as decision-making support system, based on three-dimensional modeling on the basis of hydrodynamic models of rivers and LiDAR data and satellite imageries.

Systems and services with a similar purpose are also being developed in the Russian Federation. These include:

Ground-based and satellite-based flood monitoring, forecasting, and early warning system "GIS Amur" [24], which is used for:

- Collection, systematization, visualization, and analysis of data of hydrometeorological observations, calculations, and forecasts
- Automatic release of hydrological forecasts for Amur River basin
- Aggregation of ground-based, satellite, and model information for the analysis of hydrological situation in the Amur River basin
- Real-time delivery of actual and prognostic information about hydrological situation in Amur River basin for customers concerned

By using Research Center for Earth Operative Monitoring (NTs OMZ) [25] archival materials of detailed resolution satellite imageries, it is possible to update various thematic maps of scale smaller than 1:10,000, create digital ortho-images with a resolution from 1 to 10 m, and generate DEM and DTM.

Today "NTsOMZ" is proposing to its customers:

- Products on natural flood monitoring, from thematic processing of satellite data to exploiting information about hydrological facilities, ice conditions, and others
- Thematic geospatial data sets to verify patterns of flooding on the basis of remote sensing materials (satellites data from Landsat 7, Meteor-M, and Resurs-P)

The running of flood monitoring and forecasting systems is based on models and methods that can be conditionally combined into several large groups.

Satellite monitoring has been one of the most important methods so far. It includes the use of both radar [26–32] and optical [33] images. Although the benefits of this approach are widely known, it is far from being flawless from the decision-making point of view. Most notably, it provides only post-event mapping and the boundaries of an inundation as well as post-event damage analysis.

The second group of well-known methods are medium-term and long-term forecasting for a period of several weeks to several months. These methods are based on hydrological modeling, and they require large amounts of initial data, such as the characteristics of the snow cover, soil properties, meteorological parameters, etc. [34]. Naturally, medium-term and long-term forecasting give valuable information which allows analyzing possible scenarios for probable flooding in the following few months. Nevertheless, the low reliability of the initial data and the large time interval of forecasting do not provide accurate estimates of the flooding time at each specific point of the river valley.

So, the first basic question is whether it is possible to find a compromise between the quality of the initial data and the duration of the forecast interval, on the one hand, and the accuracy of forecasting on the other hand. In the investigation presented here, this balance has been found by shifting to the short-term (operational) forecasting of river floods. In this case the forecast is performed in advance for 12–48 h, and initially it starts only when the water level increases.

The second basic question is how can we combine, integrate, and synchronize all necessary remote sensing data, other heterogeneous data, information, and models to predict the zones and depths of flooding and at the same time perform the necessary verification and validation procedures to ensure the required reliability of the results and provide the results to users in a timely and convenient form.

In previous papers [35, 36], it was described how satellite monitoring and traditional hydrological forecasting approaches based on river physical processes modeling can be enhanced by integrating different models and technologies – such as input data clustering and filtering, digital mapping of river valleys relief, data crowdsourcing, hydrological and hydrodynamic modeling, inundation visualization, duly warning of stakeholders, and also techniques for flooding scenario generation and comparison. The results described in that papers have been continuing the investigations [37–42] on the use of heterogeneous data for flood monitoring and forecasting. With that, they appear to be more universal in terms of types of data integrated in the joint system.

The need to develop systems that combine heterogeneous data and models has already been noted by a number of authors [43]. To date, however, experience has yet to be gained in development and practical application of such systems. So far there are no uniform principles of such systems architecture designing for the integration of heterogeneous data (including RS data). Experience of information technology developments and applications in other subject areas demonstrate that the approach based on service-oriented architecture (SOA) [44] is an acceptable one for design of distributed integrating systems.

The principles of software implementation of a flood forecasting system based on SOA had been previously proposed, and the first successful case study using in this system one of the possible hydrodynamic models – LISFLOOD [9] – had been earlier described [35, 36].

But it should be noted that high-quality forecasts need fundamentally not one but a complex ensemble of hydrodynamic models to choose model types and parameters depending on the specifics of the riverbed section simulation, the nature of water movement along the riverbed and floodplain, the initial data composition, and a number of other factors. The multimodel approach can significantly improve the forecasting quality but implies the development and implementation of new methods of automatic (or interactive) model selection and the model parameters adjustment.

To date, some individual examples of the use of the multimodel approach are already known, in particular, the use in running order of the 1D hydrodynamics and 2D flood spreading models [45]. These approaches call for further development of sharing hydrodynamic models and remote sensing data, as well as automation of the model choice.

So the third basic question is how can we choose and adjust the model automatically. In this chapter, a method and algorithms for the multimodel approach are presented in close correlation with the previous results related to operational flood forecasting system (OFFS) creation and testing.

2 Hydrodynamic Models

Before describing satellite image application, let us consider the basic components of a flood forecasting system, which specify requirements for an information support for modeling.

Firstly, the basic components of the OFFS are the hydrodynamic models.

Thus far, various countries have developed and implemented a large number of such models: the Russian software packages Flood, River, and STREAM-2D, developed by V.V. Belikov, A.N. Militeev, and others [46]; Mike 21 of the Danish Hydraulic Institute [47]; Delft 3D of the Deltares Institute in Delft, the Netherlands [48]; HEC_RAS of the US Army Corps of Engineers [49]; and others. The models mainly differ in the methods of schematization of the computational domain (triangular, tetragonal, and mixed meshes), computational schemes and methods (finite differences and finite elements), and sets of additional blocks (admixture dissemination, load transport, etc.).

Regardless of the specific software implementation, mathematical models of water flow movement are mostly based on the set of Saint-Venant equations [46]. When writing the set of Saint-Venant equations, the water flow can be represented both in the one-dimensional (flow characteristics are averaged by cross section) and in the two-dimensional (averaged by depth) schematizations. These equations are widely used in computing the hydraulics of open channels [50]. They consider the main forces operating on a stream with a free surface (gravity, friction, pressure, and inertia; Coriolis's force and wind influence can be considered in addition) and the three-dimensional orography of the land surface.

As a basis for modeling water movement, one- and two-dimensional models of the STREAM-2D software suite (an upgraded version of the River software package) [46] were selected. This suite is based on approaches of irregular hybrid computational meshes and original methods of interpolation [51]. It has demonstrated high efficiency in solving various problems concerned with the analysis of water flow dynamics and inundations caused by both natural and anthropic factors including ice jams.

Detailed information about the topography of the region and the riverbed relief is necessary for model setup. Water discharges and water levels at the modeling area boundaries are utilized as the model input. As a result of modeling, one can determine the flooded areas and obtain a spatial distribution of mean vertical flow velocities, water surface levels, and depths at any point of the channel and the inundated floodplain.

3 Models Selection

The design of OFFS implies, as was noted above, that no universal model exists to describe flood development processes at river sections that are different in length and configuration. When choosing hydrodynamic models to solve flood forecasting problems, it is advisable to implement the multimodel approach: depending on the length of the monitored section and the presence of initial data, it is possible to choose between one-dimensional hydrodynamic models for river valley sections of 100–1000 km long and two-dimensional models for sections shorter than 100 km with wide beds and floodplains, complex configurations, and with various structures located inside them.

Combined (hybrid, complex, multiscale) computing using one-dimensional and two-dimensional models can yield a significant effect in effort and fund economy when investigating and monitoring lengthy river areas.

Therefore, a principal issue is to form a mechanism of choosing the most adequate model for specific conditions. When building operational flood forecasting systems, this technology is implemented through the creation of an intelligent interface, which unites heterogeneous in situ and space-based data and expert knowledge to be used later for forecasting.

The intelligent interface's main purpose is to choose and adapt a specific model to compute water propagation zones and flood depths on the basis of contextual information (baseline data accuracy, flood dynamics, operational result efficiency, etc.).

Thus far, the theory, methods, and technologies for the design and implementation of various model classes (conceptual, mathematical, logical–algebraic, logical–linguistic, and mixed) of complex objects and complex processes have been developed [52]. At the same time, quality assessment of models, their systematization by various classes, and substantiations of the synthesis of new or the choice from among the existing most preferable models for solving specific applied problems remain practically unsolved under the current conditions. The topicality of the above problems is amplified when the object analyzed is described not by one model but by a polymodel complex, which includes heterogeneous and combined models, to assess each of which a special system of indicators is necessary [53]. Assessment becomes even more complex if it is necessary to take into account the time factor, which affects the model quality. This primarily concerns original objects with significant structural dynamics under the action of various (objective, subjective, internal, external, etc.) causes [53]. Then, the models used should be adapted to changing conditions to preserve their accuracy. For this, it is necessary well in advance, at the stage of model synthesis, to provide the model's structure and parameters with additional elements that, when used directly, will make it possible to control the model quality, reducing the sensitivity of its corresponding quality indicators to changes in the composition, structure, and content of initial data. Within the theory of model qualimetry and polymodel complexes, which the authors of this article are developing [52], several technologies of their structural–parametric adaptation have been designed. One of them, applicable to operational flood forecasting, is given in Fig. 1. When building OFFS, this technology is implemented through the creation of an intelligent interface, which unites heterogeneous in situ and space-based data and expert knowledge to be used later for forecasting.

The diagram given in Fig. 1 shows that, based on the results of modeling quality assessment in block 6, it is possible to control the selection of models and their structures (blocks 9 and 10) as well as adjust the model parameters (block 8). The initial information for control is based on the results of comparing the actual situation on the flooded areas that comes from block 4 with the results of simulation and forecasting from block 5. In practice, satellite images of flooded areas are one of the main data sources that come from block 4 for assessing the quality of modeling.

The intelligent interface main purpose is to choose and adapt a specific model to simulate water propagation and flood depths on the basis of contextual information (initial data accuracy, flood dynamics, operational result efficiency, etc.). The methods and algorithms that are currently being developed intensively within the theory of evolutionary modeling have been chosen as the scientific basis for the adaptation

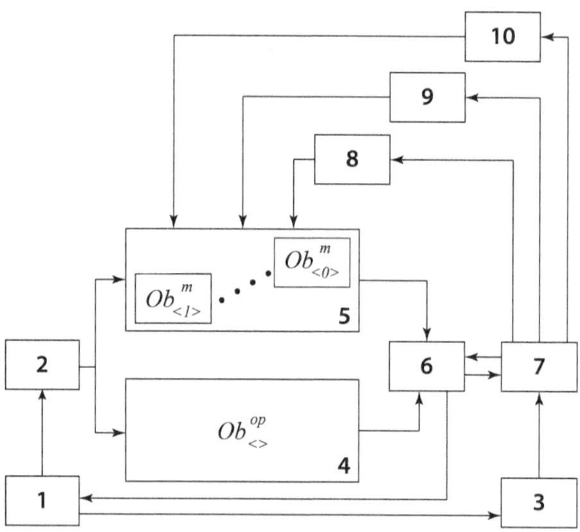

Fig. 1 Generalized technology to assess hydrodynamic models and control their quality: (*1*) formation of targets and requirements on operational flood forecasting results, (*2*) input data (scenarios) acquisition (formation), (*3*) formation of simulation objectives, (*4*) simulated system (object–original $\mathrm{Ob}^{\mathrm{op}}_{<>}$), (*5*) models ($\mathrm{Ob}^{m}_{<\theta>}$) of the analyzed system $\mathrm{Ob}^{\mathrm{op}}_{<>}$, (*6*) assessing the quality of the model (polymodel complexes), (*7*) model quality control, (*8*) model parameters control, (*9*) model structures control, and (*10*) changing the model description concept (changing boundaries within the object–medium pair)

of hydrodynamic models [54]. Overall, the creation of an intelligent interface ensures a synergetic effect from the joint processing and use of heterogeneous in situ data and the implementation of a model complex to assess possible flood areas.

4 OFFS Structure and Operation Algorithm

The real-time OFFS has been created with the approach described above, and it is based on the integrated use of in situ and remotely sensed data. This system allows operational forecasting of the river water levels, discharges, inundation areas, and depths of flooding and provides prior notification for the citizens in emergency situations at a geoportal and/or by using mobile devices (smartphones, tablets, computers). The system contains the following main components (Fig. 2):

- The input data collection and preprocessing components (blue color) provide input data collection (e.g., satellite images, online hydrological data from meteorological stations, a digital terrain model) and preprocessing (i.e., primarily image processing, information filtering, and information fusion).
- Automatic modeling component allows forecasting of water levels, calculation and forecasting of the water discharge, depths of flooding, and also automatic modeling of the water spread using hydrodynamic models.
- Post-processing, publishing, and visualization components provide post-processing of modeling results, output data storage in a database, flood map vectorization and visualization of inundation areas.

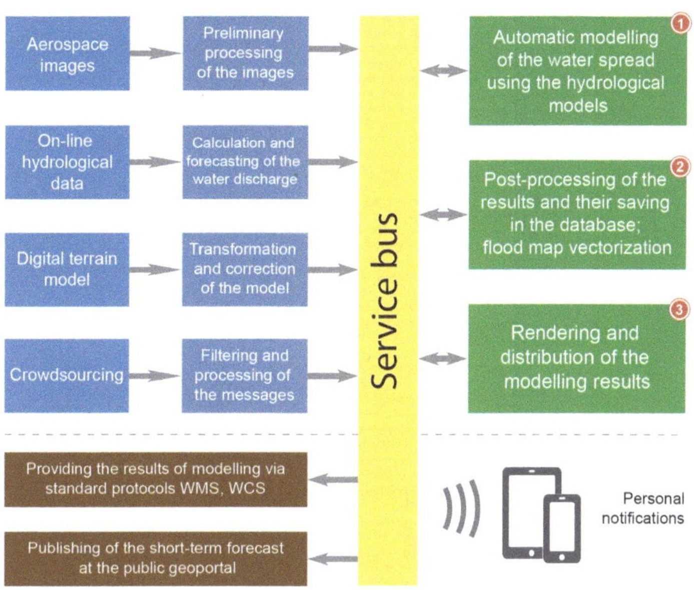

Fig. 2 The OFFS structure

Fig. 3 Analysis of inundation areas by the timeline slider

- Rendering and distribution components provide dissemination of the modeling results to the external software, publishing short-term forecasts at the public Geoportal, and automatic notification of the local citizens and organizations using web services and mobile personal devices.

The input data collection and preprocessing components are aimed at effectively integrating all the available data from various geographically distributed data sources with analytical capabilities to enhance the reliability of the monitoring information and the speed at which it becomes available to decision-makers. Another important aspect is to provide the user with the most accurate spatial and temporal resolutions of data, models, and tools, according to the actual tasks to be solved. Consequently, assessment at different geographic scales should take advantage of different data sources. The component implements automated data fusion allowing essential measurements and information to be combined to provide knowledge of sufficient richness and integrity so that decisions may be formulated and executed autonomously. The OFFS architecture and functionality are universal regarding the input data type diversity, density, and speed.

The proposed system operates automatically and provides flood forecasts for the following 12–48 h with hourly outlines of the potential flooded zones, objects, and a water depth map, both available via the standard protocols WMS and WCS. The flood forecasting results are provided as a remote web service. Moreover, the users are not required to have specific knowledge in modeling and simulation or to have programming skills.

Two additional useful functions are implemented in the OFFS. First, every user can move the timeline slider and see the inundated area 12 h in advance (Fig. 3); all the complexity is hidden from the user. The second additional function also allows the user to see the flood area for every possible water level (Fig. 4). Finally, automatic generation and analysis of flooding scenarios will allow the analysis of the river flood dynamics and to evaluate their potential effects in the near future to support preventive actions to mitigate the impacts of the floods.

Thus, the whole short-term flood forecasting algorithm includes the following main steps in the case of the multimodel approach: (1) online hydrological data

Fig. 4 Analysis of flood possible scenarios by the vertical slider

on the river water levels are obtained from a hydrological station; (2) the observed data are processed, by application of a smoothing with trend extraction, in order to predict the river water levels in the short term; (3) if necessary, a symbolic regression-based forecasting model is built which provides water flow discharge forecasts; (4) water flow discharge data, accompanied by river terrain information, are processed by an adaptive hydrodynamic model which simulates water flowing routes and thus provides forecasts on water-covered territories in the river basin, choice of the hydrodynamic model and its adaptation on the basis of integrated usage of heterogeneous data comprising remote sensing data and river terrain model, and execution of calculations; (5) visualization software, used for geo-presentation of the river simulation results of water flows and inundation areas; and (6) archival of data and images, satellite images of the current situation and data crowdsourcing through a geoportal, used to validate and calibrate the hydrodynamic model.

5 Application of Remote Sensing Data

Taking into account the vastness of river floodplains and the typical shortage of in-situ data, the implementation of the flood forecasting systems implies extensive use of Earth remote sensing (RS) data. In many cases, RS data are the basis for flood monitoring. For example, when no gauging stations are available, gauging stations fail or unforeseen events (such as dike breaches) happen. Georeferenced and classified satellite data can provide information on flood extent (area) as well as water levels for large flooded areas [32].

The inclusion of RS data in the flood forecasting systems compensates for the shortage of in situ data and allows for additional adaptation of the models used. We may single out the following vectors for implementing RS data in the OFFS design:

Updating cartographic information about the existing infrastructure in river valleys – the boundaries of settlements in the flood zone, roads, and line structures that affect the flow structure in floodplains and riverbed configurations. To this end, high-resolution optical satellite images, obtained, for example, from satellites such as Resurs-P, GeoEye, and WorldView, or from drones, during the summer–autumn low water period are used.

Acquisition of information about river valley terrain. To this end, digital terrain models (DTMs) based on highly detailed space and air surveys appear promising. At present, the main possibilities for obtaining more accurate and usable terrain models are related to the programs of radar spacecraft TerraSAR-X and TanDEM-X to create the Earth's DTMs with height resolutions of about 2 m, as well as drone-based surveys or laser scans, which yield a centimeter accuracy during DTM plotting.

Calibration and verification of the hydrodynamic models based on information about the boundaries of flooded territories from space images [55] and crowdsourcing results. The calibration parameters of hydrodynamic models are data describing the roughness of the underlying surfaces which determines hydraulic resistance and which is characterized, as a rule, by the roughness coefficient. Setting the fields of roughness coefficients as a first approximation is based on decoding and highlighting areas with various plant covers and microreliefs, as well as on selecting coefficient values for them [56]. To this end, both medium- and high-resolution images are applicable. Usually, five or six main types of underlying surfaces (beds, plowed land, meadows, forests, and urban development) are singled out. Automated data processing methods and technologies are being developed and implemented successfully to decode these surface types [57, 58].

Note that maximally accurate calculation results require integrated processing of the various data types obtained during in situ and aerospace measurements.

6 Short-Term Forecasting System Software

It follows from the previously described OFFS composition that the architecture of the system should ensure flexible interaction of many components, including the existing and potential software modules based on hydrodynamic models, input data processing modules, forecast result visualization modules, control modules, etc. In this respect, it is desirable to build an OFFS on the basis of a service-oriented architecture [36, 59].

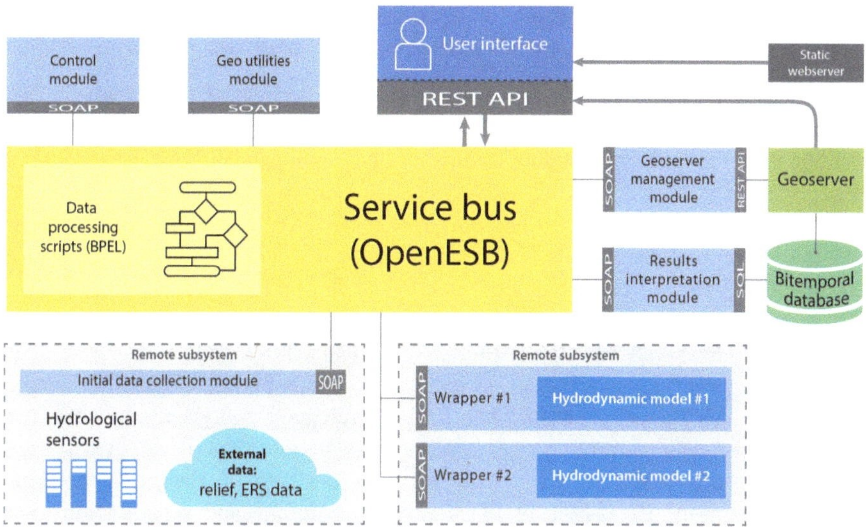

Fig. 5 Structure of the short-term forecasting system software

A software structure for the practical implementation of the described multimodal approach is presented in Fig. 5.

The service bus with intelligent interface functions is the core of the system. It is implemented as open source software – OpenESB. OpenESB has an embedded interpreter for the business processes language BPEL that executes the scenarios necessary for the functioning of an OFFS. The service bus provides the starting point for the modules intended for the collection, processing, storage, and dissemination of information, according to the scenario. Each module is a web service with a standard SOAP [44, 60] interface.

The control module performs starting and stopping of automatic calculation cycles, single calculations, and a number of auxiliary operations.

The calculation module of the intelligent interface synthesizes the particular structure of the OFFS on the basis of the initial data of the hydrological conditions.

The initial data collection module inquires of the hydrological sensors and provides the service bus with normalized data. This allows abstraction from the information exchange scheme with specific sensor manufacturers.

The wrappers of the hydrodynamic models provide the formation of files with initial data, starting calculations, performance monitoring, and collecting and sending the calculation results to the service bus.

The results interpretation module performs several stages of processing:

1. Creating the depth contours from the matrix of points;
2. Recording the data on the depth intervals and the flooded areas into the attribute table;
3. Simplification of geometry;

4. Generalization and smoothing of contours to restore the natural borders of flooded areas and improve the visualization of the simulation results.

In addition to the standard python language libraries of data analysis and visualization, such as "scipy," "numpy," and "matplotlib," the results interpretation module uses the open spatial data libraries: "ogr," "geopandas," "shapely" (vector data processing), "osr" (working with coordinate systems), matplotlib.tri.tricontourf (construction of the Delaunay triangulation elements irregular network on the basis of a field of points), matplotlib.pyplot.tricontourf (calculation of the contours points corresponding to predetermined depth intervals based on triangulation).

The GeoServer management module allows to put the automatically generated data on a GeoServer. A variety of solutions can be used as a geoserver – both open source (GeoServer, MapServer, QGIS mapserver) and commercial (ArcGIS Server, ArcIMS, GeoMedia, Oracle MapViewer). The developed prototype used GeoServer. GeoServer is managed through a networking software interface (REST API).

In the proposed system, every user receives the data on their own personal device – a smartphone, tablet, or PC. Powerful searching and visualization tools provide the users with all the necessary data on the areas and depths of inundation.

7 Case Studies and Discussions

OFFS testing in a multimodel mode was performed on the Russian rivers: Oka River (near Ryazan city), Northern Dvina River along its 100 km long course (from Veliky Ustyug to Kotlas), and Protva River.

As an example of the methodology described above, we describe in more detail the case study of Veliky Ustyug on the Northern Dvina River. The town of Veliky Ustyug, with a population of about 30,000, is situated in the northern part of European Russia at the confluence of the Sukhona and the Yug into the Northern Dvina River, which originates from their junction. Veliky Ustyug is an ancient Russian city, founded in the thirteenth century. The city was flooded repeatedly during all its history. The recurrent floods are due to snow melting and ice jams. Ice jams may cause more than 2 m additional water level rise. The last large flood was observed in May 1998, when more than half of the city and nearby territories were inundated. Significant floods took place also in April 2013 and in April 2016.

As a basis for modeling water movement, one- and two-dimensional models of the STREAM-2D software suite (an upgraded version of the River software package) [46] were selected.

Detailed information on the floodplains topography, with scale 1:25,000 and 1:10,000, was used for the STREAM-2D model setup. Data of the riverbed relief and water surface slopes were obtained from detailed field surveys organized by the lab of soil erosion and river channel processes of the Geography Faculty of Lomonosov Moscow State University.

The discretization of the modeling area in STREAM-2D was obtained by the construction of an irregular hybrid computational mesh. Curvilinear quadrangular grid units with a spatial resolution from about 10×30 to 40×100 m for the river channels and linear constructions (such as roads, levees) on the floodplains were used. Other territories were covered by triangular grid cells with irregular spatial resolution from 50 to 300 m, depending on the necessary level of detail and available topographic information. All floodplain topography and river channel relief data were interpolated onto the centers of grid cells.

Data of measured flow velocities and water surface slopes obtained during the field campaigns were implemented for model calibration and verification as a first step. Water discharges and water levels at the gauging sites were utilized as the model input for high flood periods. Water levels for gauges located in the city of Veliky Ustyug on the Sukhona River were used as control points for model calibration and validation. Periods with ice jams for Veliky Ustyug were taken into account using a rather simple approach based on increasing the roughness and decreasing the water depths for places of ice jams, where the specific values of these parameters were obtained during additional model calibration and verification for ice jam periods.

The OFFS automatically performed the preparation and preprocessing of the initial data for the hydrodynamic models, namely, spatial information on the territory topography, the distribution of the underlying surface roughness, the time series of the water discharge, and the levels at the input and output boundaries.

Further, in the OFFS, the visualization of the modeling results was carried out.

During testing it was found that the use of a one-dimensional model is expedient for the river's elongated sections when their length is an order of magnitude or more greater than the width of the potential flood area. A two-dimensional model requires more detailed information on the river valleys morphometry, especially on the relief of potential flood territory. As a result, it allows getting a forecasted distribution of flow velocities and of water surface levels and depth within the calculation domain.

A significant effect can be achieved by a joint (hybrid, multiscale) calculation of the one-dimensional and two-dimensional models. Here the entire elongated section of the river is modeled by a one-dimensional system of equations and part of it by a two-dimensional or bilayer two-dimensional system.

The satellite images for a study area allowed performing additional calibration and validation of the model through comparison of simulated inundated areas with the outlines of flooded zones at the time of peak flow. In satellite images for the periods of high flood, Resource-01 for the flood of 1998, SPOT-5 for the flood of 2012, and Resurs-P for the flood of 2016 were used for the model calibration and validation, respectively (Table 1). Remote sensing data were provided by ScanEx Research and Development Center [61] and Research Center for Earth Operative Monitoring [62].

The results of multimodal forecasting have demonstrated a good correspondence of the calculated and measured water levels (the difference is not more than about 30 cm) and flood area boundaries. Analyses of the modeling results have also shown good correspondence between flood-modeled areas and contours derived

Table 1 Results of STREAM-2D model calibration and validation using remote sensing data for Veliky Ustyug key study area

Date	Satellite image	Spatial resolution, m	Discharges of the rivers at their confluence (m^3/s)		Difference between simulated and observed water levels at gauging station (cm)	Relative error of modeled flooded area in comparison with remote sensing data (%)
			Sukhona	Ug		
11.05.1998	Resource-01	30	4210	2780	−1	7.9
05.05.2012	SPOT-5	10	3200	3900	31	8.5
18.04.2016	Resurs-P	2	3650	3300	15	7,0

from satellite imageries. The difference does not exceed 10% (Fig. 6). The figure shows that during the simulation, the contours and depths of flooding, as well as flooded infrastructure objects (red color), are determined and visualized.

Thus, the approach to flood forecasting proposed in this paper, as well as OFFS realized on the basis of this approach, provides high accuracy of short-term forecasting. This is achieved, on the one hand, by applying the most accurate model for each specific section of the river bed and, on the other hand, through the integrated use of heterogeneous information to select and adjust the model parameters. An important role among these data is played by RS data, which provide the calibration of mathematical hydrodynamic models. In fact, the proposed version of the system construction shows how RS data can be embedded in the functioning of flood forecasting systems in practice, and thus possibilities of using RS data for this class of problems may be significantly expanded.

During steps of testing, the approach to OFFS creation on the basis of SOA was proved to be promising. It provides high system versatility and the possibility of its geographically distributed construction, since individual modules of the system are implemented as web services with data exchange between them by standard protocols. Convenient visualization of results, notification services, and an automatic mode of operation provide an opportunity for wide practical application of the described system.

The next steps in the development of OFFS are connected with the use of a wider set of models and also with automation of RS data selection for modeling, in order to achieve a correspondence between the parameters of satellite images (spatial resolution, etc.), digital elevation models, and the characteristics of modeling accuracy.

Acknowledgments The research described in this chapter is partially supported by the Russian Foundation for Basic Research (grants 15-07-08391, 15-08-08459, 16-07-00779, 16-08-00510, 16-08-01277, 16-29-09482-ofi-i, 16-07-00925, 17-08-00797, 17-06-00108, 17-01-00139, 17-20-01214), by the Russian Research Foundation (grants 16-19-00199, 17-11-01254), by ITMO

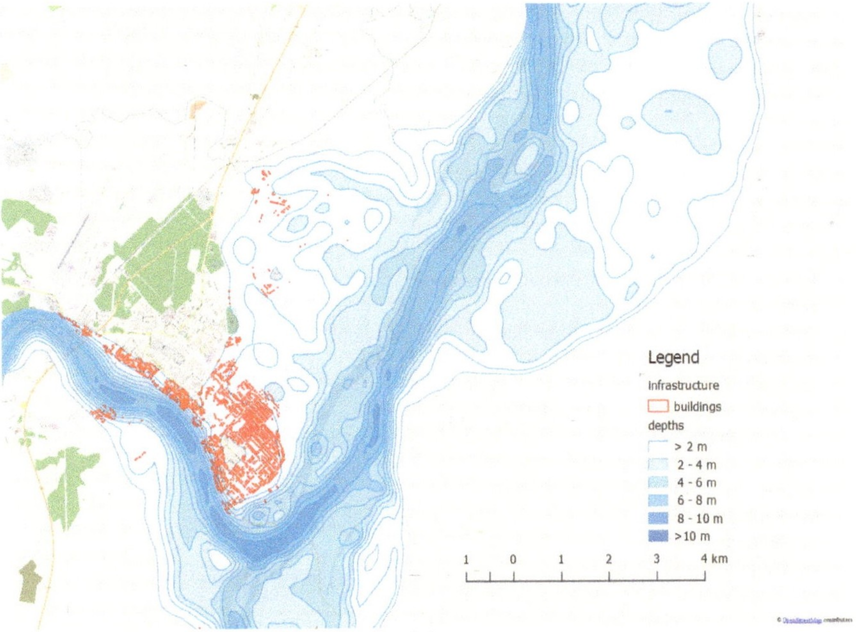

(a)

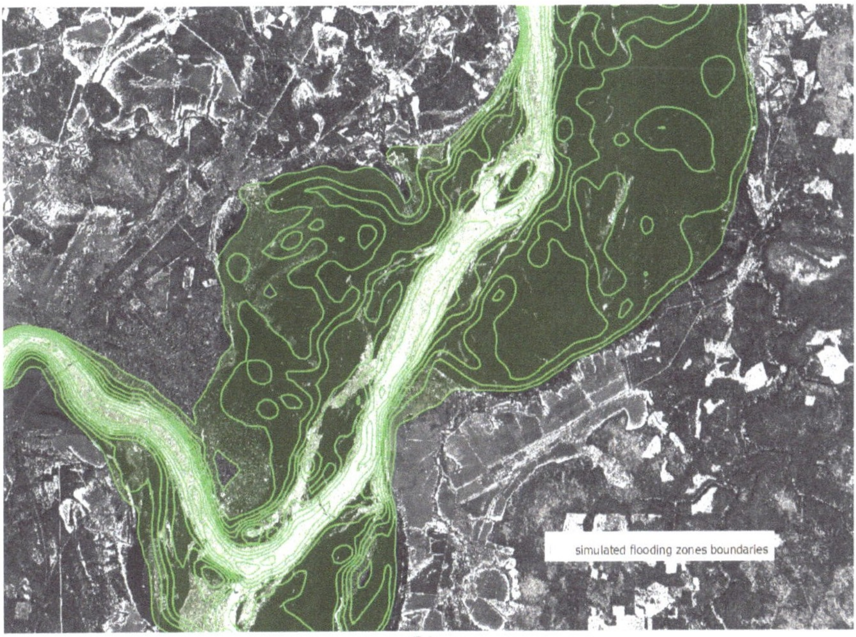

(b)

Fig. 6 Comparison of the modeling results (**a**) with the actual outlines (**b**) of inundation areas (18.04.2016, Northern Dvina River)

University's grant 074-U01, project 6.1.1 (Peter the Great St. Petersburg Polytechnic University) supported by Government of Russian Federation, Program STC of Union State "Monitoring-SG" (project 1.4.1-1), state order of the Ministry of Education and Science of the Russian Federation №2.3135.2017/K, and state research 0073–2014–0009 and 0073–2015–0007.

References

1. Transboundary Flood Risk Management in the Unece Region. United Nations New York and Geneva (2009)
2. Porfiriev, B.N.: Economic consequences of the 2013 catastrophic flood in the Far East. Herald Russ. Acad. Sci. **85**(40), pp. 40–48 (2015)
3. Alekseevskii, N.I., Frolova, N.L., Khristoforov, A.V.: Monitoring Hydrological Processes and Improving Water Management Safety. Izd. Mos. Gos. Univ, Moscow (2011) [in Russian]
4. Vasil'ev, O.F.: Designing systems of operational freshet and high water prediction. Herald Russ. Acad. Sci. **82**(129), pp. 129–133 (2012)
5. http://waterwatch.usgs.gov [Accessed 6 April 2017]
6. MIKE Flood Watch, https://www.mikepoweredbydhi.com/products/mikeflood [Accessed 6 April 2017]
7. DHI-Water forecast, http://www.waterforecast.com [Accessed 6 April 2017]
8. EFAS, http://www.efas.eu [Accessed 6 April 2017]
9. LISPFLOOD-FP, University of Bristol, School of Geographical Sciences, Hydrology Group, www.bristol.ac.uk/geography/research/hydrology/models/lisflood/ (Accessed 6 Apr 2017)
10. FEWS, https://www.deltares.nl/en/software/floodforecasting-system-delft-fews-2/ [Accessed 6 April 2017]
11. https://www.mikepoweredbydhi.com/products/mike-11 [Accessed 6 April 2017]
12. https://www.mikepoweredbydhi.com/products/mike-21 [Accessed 6 April 2017]
13. https://oss.deltares.nl/web/delft3d/home [Accessed 6 April 2017]
14. http://www.hec.usace.army.mil/software/hec-ras/ [Accessed 6 April 2017]
15. https://www.flo-2d.com/ [Accessed 6 April 2017]
16. http://www.opentelemac.org/index.php/presentation [Accessed 6 April 2017]
17. WRF, www.wrf-model.org [Accessed 6 April 2017]
18. http://www.meteorf.ru [Accessed 6 April 2017]
19. https://lance-modis.eosdis.nasa.gov [Accessed 6 April 2017]
20. http://emergency.copernicus.eu/mapping/ems/service-overview [Accessed 6 April 2017]
21. https://www.digitalglobe.com/products/advancedelevation-series [Accessed 6 April 2017]
22. GLCF, www.landcover.org [Accessed 6 April 2017]
23. https://nrsc.gov.in/floods [Accessed 6 April 2017]
24. Frolov, A.V., Asmus, V.V., Zatyagalova, V.V., Krovotyntsev, V.A., Borshch, S.V., Vil'fand, R.M., Zhabina, I.I., Kudryavtseva, O.I., Leont'eva, E.A., Simonov, Y.A., Stepanov, Y.A.: GIS-Amur system of flood monitoring, forecasting, and early warning. Russian Meteorol. Hydrol. **41**(3), 157–169 (2016)
25. http://eng.ntsomz.ru/ [Accessed 6 April 2017]
26. D'Addabbo, A., et al.: A Bayesian network for flood detection combining SAR imagery and ancillary data. IEEE Trans. **54**(6), pp. 3612–3625 (2016)
27. Mason, D., Davenport, I., Neal, J., Schumann, G., Bates, P.D.: Near real-time flood detection in urban and rural areas using high resolution synthetic aperture radar images. IEEE Trans. Geosci. Remote Sens. **50**(8), 3041–3052 (2012)
28. Pierdicca, N., Pulvirenti, L., Chini, M., Guerriero, L., Candela, L.: Observing floods from space: experience gained from COSMO-SkyMed observations. Acta Astron. **84**, 122–133 (2013)

29. Romeiser, R., Runge, H., Suchandt, S., Sprenger, J., Weilbeer, H., Sohrmann, A., Stammer, D.: Current measurements in rivers by spaceborne along-track InSAR. IEEE Trans. Geosci. Remote Sens. **45**(12), 4019–4031 (2007)
30. Hahmann T.. Wessel B..: Surface Water Body Detection in High-Resolution TerraSAR-X Data using Active Contour Models. In: Proceedings of the 8th European Conference on Synthetic Aperture Radar (EUSAR 2010), Aachen, Germany. VDE Verlag GmbH (June 7–10, 2010)
31. Floyd, L., Prakash, A., Meyer, F.J., Gens, R., Liljedahl, A.: Using synthetic aperture radar to define spring breakup on the Kuparuk River, Northern Alaska. ARCTIC. **67**(4), 462–471 (2014)
32. Barneveld H.J., Silander J.T., Sane M., Malnes E..: Application of satellite data for improved flood forecasting and mapping. In:4th International Symposium on Flood Defence:Managing Flood Risk, Reliability and Vulnerability Toronto, Ontario, Canada, p. 77–1–77-8 (May 6–8 2008)
33. Wang, Y., Colby, J., Mulcahy, K.: An efficient method for mapping flood extent in a coastal floodplain using Landsat TM and DEM data. In. Int. J. Remote Sens. **23**(18), 3681–3696 (2002)
34. Sokolov B.V., Zelentsov V.A., Mochalov V.F., Potryasaev S.A., Brovkina O.V.: Complex Objects Remote Sensing Monitoring and Modelling: Methodology, Technology and Practice. In: Proc. 8th EUROSIM Congress on Modelling and Simulation, 10–13 September 2013, Cardiff, Wales, United Kingdom, pp.443–447 (2013)
35. Merkuryev, Y., Merkuryeva, G., Sokolov, B., Zelentsov, V. (eds.): Information Technologies and Tools for Space-Ground Monitoring of Natural and Technological Objects, Riga, Riga Technical University (2014)
36. Boris V. Sokolov, Anton Ev. Pashchenko, Semyon Potryasaev A., Alevtina V. Ziuban, Vyacheslav A. Zelentsov.: Operational Flood Forecasting As A Web-Service. In.: Proc. 29th European Conference on Modelling and Simulation (ECMS 2015), Albena (Varna), Bulgaria. P. 364–370 (2015)
37. Schumann G., Bates P., Horritt M., Matgen P., Pappenberger F.: Progress in integration of remote sensing-derived flood extent and stage data and hydraulic models. Rev. Geophys. **47**(4), Art. no. RG4001 (2009)
38. Brivio, P.A., Colombo, R., Maggi, M., Tomasoni, R.: Integration of remote sensing data and GIS for accurate mapping of flooded areas. Int. J. Remote Sens. **23**(3), 429–441 (2002)
39. Matgen, P., Schumann, G., Henry, J., Hoffmann, L., Pfister, L.: Integration of SAR-derived river inundation areas, high-precision topographic data and a river flow model toward near real-time flood management. Int. J. Appl. Earth Observ. Geoinf. **9**(3), 247–263 (2007)
40. Loumagne C., Weisse A., Normand M., Riffard M., Quesney A., Le. Hegarat-mascle S., Alem F.: Integration of remote sensing data into hydrological models for flood forecasting. In: Remote Sensing and Hydrology 2000 (Proceedings of a symposium held at Santa Fe, New Mexico, USA). IAHS Publ. no. 267, 2001, p. 592–594 (Apr 2000)
41. Hartanto, I.M., Almeida, C., Alexandridis, T.K., Weynants, M., Timoteo, G., Chambel-Leitao, P., Araujo, A.M.S.: Merging earth observation data, weather predictions, in-situ measurements and hydrological models for water information services. Environ. Eng. Manag. J. **14**(9), 2031–2042 (2015)
42. van Dijk, A.I.J.M., Renzullo, L.J.: Water resource monitoring systems and the role of satellite observations. Hydrol. Earth Syst. Sci. **15**, 39–55 (2011)
43. Lehmann, A., Giuliani, G., Ray, N., Rahman, K., Abbaspour, K.C., Nativi, S., Craglia, M., Cripe, D., Quevauviller, P., Beniston, M.: Reviewing innovative Earth observation solutions for filling science-policy gaps in hydrology. J. Hydrol. **518**(Part B), 267–277 (2014)
44. Thomas Erl.: Next generation SOA: a concise introduction to service technology & service-orientation. In: The Prentice Hall Service Technology Series from Thomas Erl). M.: Prentice Hall. Upper Saddle River, New Jersey, 234 p. (2014)

45. Emma, B., Danie, B., Michael, C., de Annemargreet, L., Leonore B., Ferdinand, D., Dirk, E., de Karin, B., Albrecht, W., Caroline, H., Joost, B.: Methods and tools to support real time risk-based flood forecasting – a UK pilot application. In: FLOODrisk 2016 – 3rd European Conference on Flood Risk Management, E3S Web of Conferences 7, Lyon, France, 18019 (2016)
46. Belikov V.V., Krylenko I.N., Alabyan A.M., Sazonov A.A., Glotko A.V.: Two-dimensional hydrodynamic flood modelling for populated valley areas of Russian rivers. In: Proceedings International Association of Hydrological Sciences, Prague, Czech Republic, pp. 69–74 (2015)
47. Skotner C. et al.: MIKE FLOOD WATCH - managing real-time forecasting, www.researchgate.net/publication/242572811_MIKE_FLOOD_WATCH_-_Managing_Real-Time_Forecasting (Accessed 6 Apr 2017)
48. Delft3D-FLOW Version 3.06 User Manual., oss.deltares.nl/web/delft3d (Accessed 6 Apr 2017)
49. HEC-RAS river analysis system User's Manual., www.hec.usace.army.mil/software/hec-ras. (Accessed 6 Apr 2017)
50. Cunge, G.A., Holly, F.M., Verway, A.: Practical Aspects of Computational River Hydraulics. Pitman Publishing LTD, London (1980)
51. Belikov, V.V., Militeev, A.N.: The two-layered mathematical model of catastrophic floods. Vych.Tekhnol. **3**, 167 (1992)
52. Sokolov, B.V., Yusupov, R.M.: Conceptual foundations of quality estimation and analysis for models and multiple-model systems. J. Comput. Syst. Sci. Int. **43**(6), 831 (2004)
53. Okhtilev, M.Y., Sokolov, B.V., Yusupov, R.M.: Intelligent Technologies for Monitoring and Controlling the Structural Dynamics of Complex Technological Objects. Nauka, Moscow (2006.) [in Russian]
54. Bukatova, I.L.: Evolutionary Modeling and Its Applications. Nauka, Moscow (1979.) [in Russian]
55. Merkuryev Y., Okhtilev M., Sokolov B., Trusina I., Zelentsov V.: Intelligent Technology for Space and Ground based Monitoring of Natural Objects in Cross-Border EU-Russia Territory. In: Proc. International Geoscience and Remote Sensing Symposium (IGARSS 2012), Munich, Germany, pp. 2759–2762 (2012)
56. Baryshnikov, N.B.: Hydraulic Resistances of Riverbeds. Izd. Ross. Gos. Gidrometeorol. Univ, St. Petersburg (2003.) [in Russian]
57. Sokolov B.V., Zelentsov V.A., Brovkina O., et al.: Complex objects remote sensing forest monitoring and modeling. In: Ed. Silhavy, R., Senkerik, R., Oplatkova, Z.K., Silhavy P., Prokopova, Z. (eds.) Modern Trends and Techniques in Computer Science: Advances in Intelligent Systems and Computing, Cham, Switzerland, Vol. 285, pp. 445–453. Springer (2014)
58. Sokolov, B.V., Zelentsov, V.A., Yusupov, R.M., Merkuryev, Y.A.: Multiple models of information fusion processes: quality definition and estimation. J. Comput. Sci. **5**(380), pp. 380–386 (2014)
59. Vasiliev Y.: SOA and WS-BPEL: Composing Service-Oriented Solution with PHP and ActiveBPEL, Packt Publishing, Birmingham, United Kingdom (2007)
60. https://www.w3.org/TR/soap/ [Accessed 6 April 2017]
61. Moscow, Russia, www.scanex.ru/ [Accessed 6 April 2017]
62. NTs OMZ, http://eng.ntsomz.ru/ [Accessed 6 April 2017]

Monitoring Flood Extent and Area Through Multisensor, Multi-temporal Remote Sensing: The Strymonas (Greece) River Flood

Alberto Refice, Annarita D'Addabbo, Francesco Paolo Lovergine, Khalid Tijani, Alberto Morea, Raffaele Nutricato, Fabio Bovenga, and Davide Oscar Nitti

Abstract Satellite monitoring of flood events at high spatial and temporal resolution is considered a difficult problem, mainly due to the lack of data with sufficient acquisition frequency and timeliness. Typically, cloudy weather conditions associated with floods obstacle the propagation of e.m. waves in the optical spectral range, forbidding acquisitions by optical sensors. This problem is not present for longer wavelengths, so that radar imaging sensors are recognized as viable solutions for long-term flood event monitoring. In selected cases, however, weather conditions may remain clear for sufficient amounts of time, enabling monitoring of the evolution of flood events through long time series of satellite images, both optical and radar. In this contribution, we present a case study of long-term integrated monitoring of a flood event which affected part of the Strymonas river basin, a transboundary river with source in Bulgaria, which flows then through Greece up to the Aegean Sea. The event started at the beginning of April 2015, due to heavy rain, and the flooded areas lasted up to the beginning of September. Due to the arid climate characterizing the area in this period of the year, weather conditions were cloud-free for most of the time interval covering the event. We collected remotely sensed data, including one high-resolution, X-band, COSMO-SkyMed and several C-band, Sentinel-1 SAR, and optical Landsat 8 images of the area. The SAR backscatter and optical NDVI maps were thresholded to obtain binary flood maps for each day. Threshold values for microwave and optical data were calibrated by comparing one SAR and one optical image acquired on the same

A. Refice (✉) • A. D'Addabbo • F.P. Lovergine • F. Bovenga
ISSIA-CNR, Bari, Italy
e-mail: refice@ba.issia.cnr.it; daddabbo@ba.issia.cnr.it; f.lovergine@ba.issia.cnr.it; bovenga@ba.issia.cnr.it

K. Tijani • A. Morea • R. Nutricato • D.O. Nitti
GAP s.r.l., c/o Dipartimento di Fisica "M. Merlin", Bari, Italy
e-mail: khalid.tijani@gapsrl.eu; eu.albertomorea@gmail.com; raffaele.nutricato@gapsrl.eu; davide.nitti@fisica.uniba.it

A. Refice et al. (eds.), *Flood Monitoring through Remote Sensing*,
Springer Remote Sensing/Photogrammetry,
https://doi.org/10.1007/978-3-319-63959-8_5

date. Results allow to draw a multi-temporal map of the flood evolution with high temporal resolution. The extension of flooded area can also be tracked in time, allowing post-flood recovery monitoring, as well as to envisage future testing of evapotranspiration/absorption models.

Keywords Multi-temporal flood monitoring • SAR-optical data integration • DEM-based flood area assessment

1 Introduction

Satellite monitoring of floodwater at high spatial and temporal resolution is considered a very important problem for several applications, including hydrology [1], hydraulic modeling [8], and water quality mapping [11]. Yet, it is a difficult problem, mainly due to the lack of data with sufficient acquisition frequency and timeliness [9]. The problem is worsened by the typically cloudy weather conditions associated with floods, which obstacle the propagation of e.m. waves in the optical spectral range, forbidding acquisitions by optical sensors. This problem is not present for longer wavelengths, so that radar imaging sensors are recognized as viable solutions for long-term flood event monitoring [5, 13, 14], with an outlook to operational implementations [18]. In selected cases, however, weather conditions may remain clear for sufficient amounts of time, enabling monitoring of the evolution of flood events through long time series of satellite images, both optical and radar. Efforts in this direction have been devoted typically toward low- to medium-resolution sensors [7], which ensure high-frequency acquisitions and thus increase the rate of cloud-free acquisitions at mid-latitudes. Other examples are long river courses, in situations where floods may be triggered significantly downstream with respect to the location where precipitation occurred, with different local weather conditions.

In this contribution, we present a case study of long-term integrated monitoring of a flood event which affected part of the Strymonas river basin, a transboundary river with source in Bulgaria, which flows then through Greece up to the Aegean Sea. The event, which affected the floodplain close to the river mouth, started at the beginning of April 2015, due to heavy rain upstream of the monitored area; flooded areas lasted for several months, with some water pools still present at the beginning of September. Due to the arid climate characterizing the area in the monitored season, weather conditions were prevalently cloud-free for most of the period covering the event. We collected a multi-temporal dataset consisting of a high-resolution, X-band, COSMO-SkyMed, C-band, Sentinel-1 SAR, and optical Landsat-8 images of the area. Results allow to draw a multi-temporal map of the flood evolution with high temporal resolution. The extension of flooded area can also be tracked in time, allowing some interesting insight into the post-flood recovery process.

2 Data and Methods

The map on the left of Fig. 1 shows the large-scale location of the investigated area. Table 1 shows the available data used. In the table, image dates are reported in temporal sequence. Note the availability of both SAR and optical data on the date of April 4, 2015, although acquired at different times. The L8 image acquired in February shows no sign of flood and has been used to derive the "normal" river course extension. Although some of the used frames do not cover entirely the interest area (red rectangle in Fig. 1), each chosen image covers the total flooded area at its respective acquisition date.

Maps with all the used data are reproduced in Fig. 2. The partial cloud cover is visible in many of the optical images. Moreover, as mentioned, some of the images do not cover completely the black rectangle. Nevertheless, it can be noticed that the flooded area boundary at each date is completely covered by the image. Absence of surface water in the uncovered scene areas can be safely assumed, since no significant rain is recorded in the days following the flood, so that, for instance, since no standing water can be seen on the southernmost "pool" in the SAR backscatter S1 image of May 22, then the L8 NDVI image of June 14, and all subsequent ones, can be safely used even though the southern part is not visible in some of them. This partial coverage is accounted for during processing by properly masking out uncovered parts from images (Fig. 3).

SAR images were absolutely calibrated. Then, they were speckle-filtered through a filter based on a nonlocal paradigm [6], which has been shown to smooth out homogeneous regions while retaining high resolution over terrain features causing sharp signal variations, such as man-made point-like or linear features, or, more appropriately, the sharp edges between water and dry areas. The speckle-filtered σ^0 images were then precisely geocoded through the range-Doppler algorithm [15]; optical images were radiometrically corrected to obtain ground reflectance values [3] from which NDVI maps were derived.

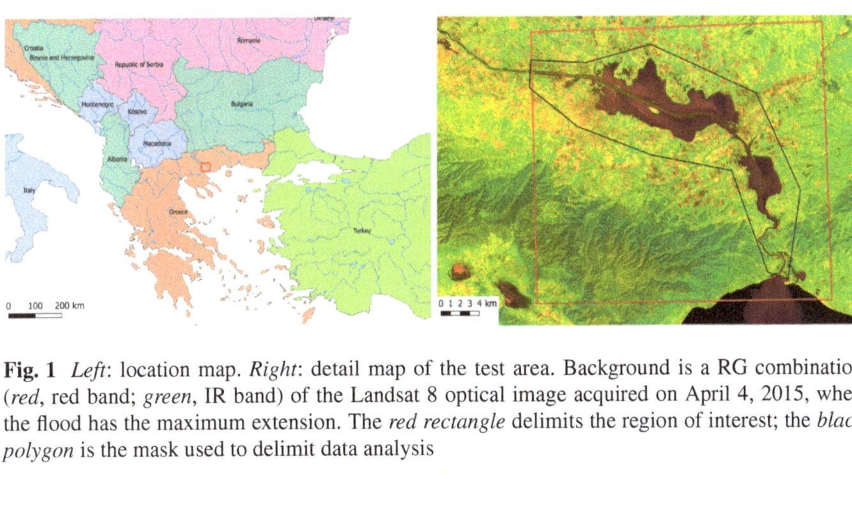

Fig. 1 *Left*: location map. *Right*: detail map of the test area. Background is a RG combination (*red*, red band; *green*, IR band) of the Landsat 8 optical image acquired on April 4, 2015, when the flood has the maximum extension. The *red rectangle* delimits the region of interest; the *black polygon* is the mask used to delimit data analysis

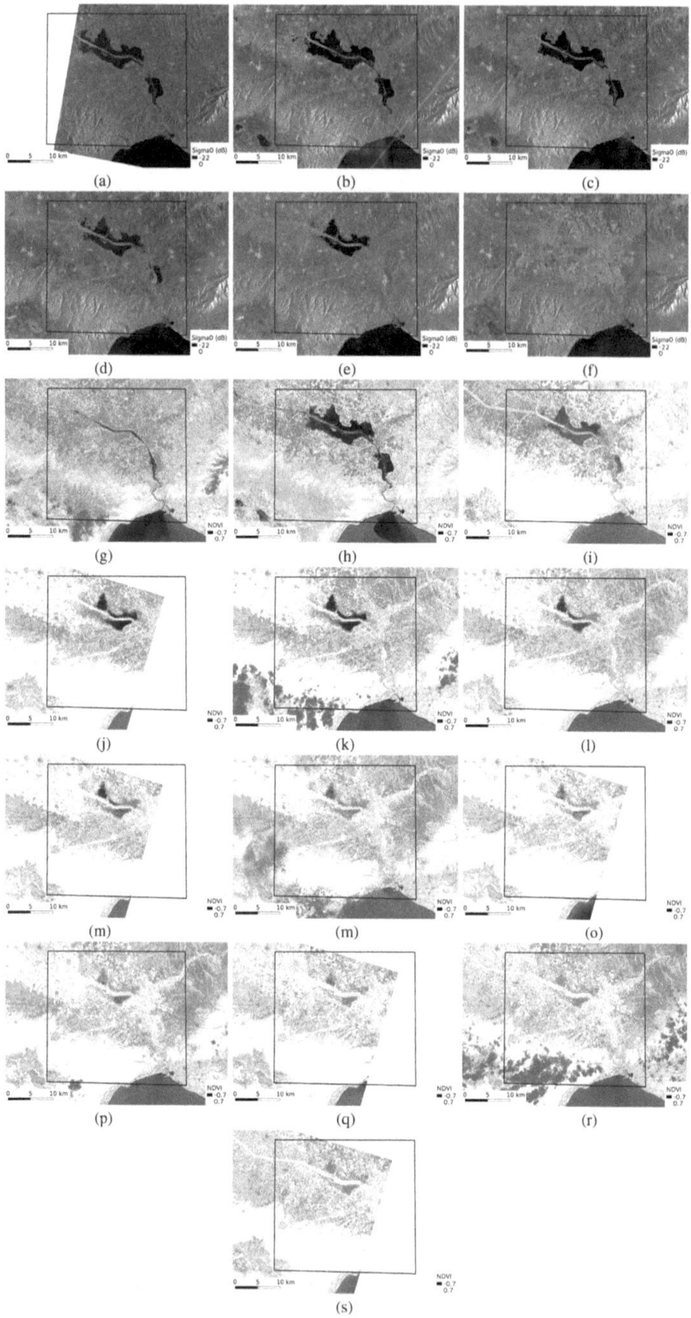

Fig. 2 Maps of the dataset used. (**a**) COSMO-SkyMed σ^0; (**b–f**) Sentinel-1 σ^0; (**g–s**) Landsat 8 NDVI. Images for each sensor are shown in temporal sequence, according to the acquisition dates reported in Table 1. The *black rectangle* shows the area of interest

Table 1 Image dataset and related acquisition dates and times

COSMO-SkyMed	Sentinel-1	Landsat 8
		15/02/2015 – 09:04:08
01/04/2015 – 16:11:07		
	04/04/2015 – 04:30:34	04/04/2015 – 09:03:42
	16/04/2015 – 04:30:23	
	28/04/2015 – 04:30:24	
		06/05/2015 – 09:03:28
	22/05/2015 – 04:30:25	
		14/06/2015 – 09:09:47*
		23/06/2015 – 09:03:39
		09/07/2015 – 09:03:49
		16/07/2015 – 09:10:04*
		25/07/2015 – 09:03:56
		01/08/2015 – 09:10:08*
		10/08/2015 – 09:03:59
		02/09/2015 – 09:10:19*
	07/09/2015 – 04:30:32	
		11/09/2015 – 09:04:14
		18/09/2015 – 09:10:28*

* denote partial coverage of the entire region of interest

Then, both SAR and optical images were resampled to a common grid with 10 m spacing in both directions. A buffer zone, visible as a black polygon in the map at the right of Fig. 1, largely encircling the maximum extent of the flooded area, was considered around the river course, and values within such area were analyzed. Figure 4 shows the normalized histograms of the various images, calculated over the buffer area. As can be seen, despite different acquisition conditions and geometries, thanks to the absolute calibration and atmospheric correction, and to the use of the buffer zone reducing the amount of data, all histograms from each given sensor type have similar behavior, mostly exhibiting a clear bimodal distribution, with a few histograms having multiple peaks.

Both SAR backscatter and optical NDVI are expected to be low over water areas. SAR data are sensitive to the roughness of the scattering interface, so that radiation impinging on a very smooth surface is expected to be reflected mostly in the specular direction, normally away from the sensor. Optical data scattered from water bodies contain usually low amounts of infrared radiation, with respect to other spectral components, so that NDVI, defined as $(\rho_{\mathrm{NIR}} - \rho_{\mathrm{R}})/(\rho_{\mathrm{NIR}} + \rho_{\mathrm{R}})$, where ρ_{NIR} and ρ_{R} are reflectance values in the near-infrared and red spectral bands, respectively, is expected to be low on water areas. A separation of water and non-water in such kinds of data can thus be obtained by simple application of a threshold. Several work have been devoted to the definition of more or less automated methods to determine such threshold values. The main problem in this case is given by the typical very different

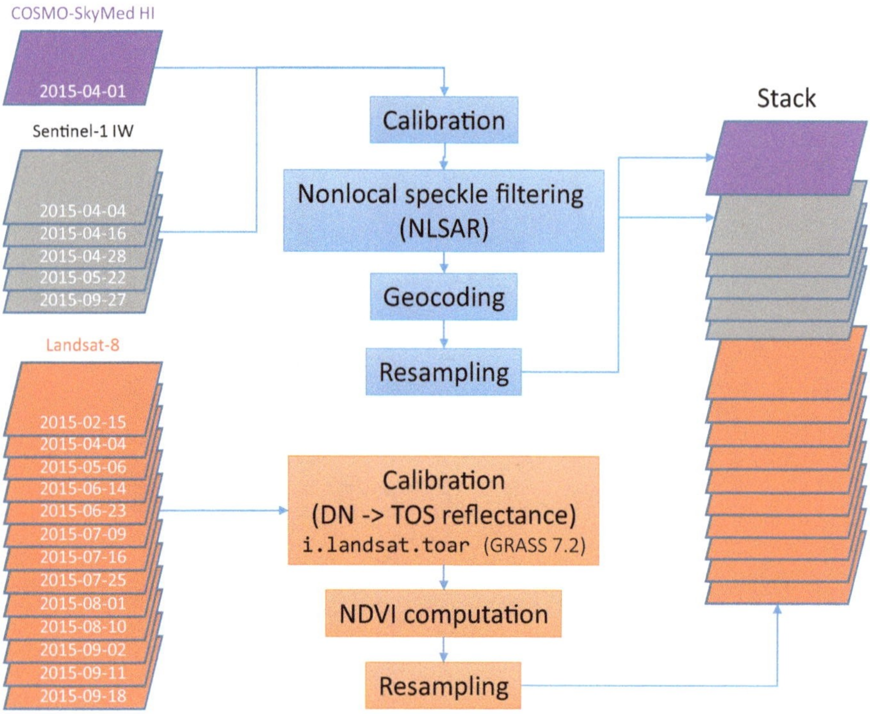

Fig. 3 Schematic diagram of the data processing steps

populations of the two pixel classes, which alters the histogram appearance, often making extremely difficult to discern a proper threshold value. Several approaches have been proposed to deal with this problem [2, 12].

In our case, the use of the buffer area encircling the maximum extent of the flood strongly limits the abovementioned problems, which may be due to different numbers of flooded and non-flooded pixels in each data sample, so that a discernible common feature of all the curves shown in Fig. 4 is a clear minimum at values around -15 dB for SAR σ^0 and around 0.1 for the NDVI. This kind of common behavior allows to assume that a single threshold value can be used (for each sensor type) to separate water-covered areas in all the images, thus greatly simplifying the data processing step.

To determine the best values for the thresholds to be used for the radar and optical data, we observe that we have both types of data acquired on the single date of April 4, 2015, at only a few hours difference. Assuming thus the ground truth to be practically the same for both these acquisitions, we seek the threshold values around the histogram minima which maximize the agreement between the flood maps obtained by the two sensors, shown in Fig. 5. In practice, we consider a grid of threshold value pairs, spanning intervals between -15 and -17 dB with a step of 0.1 dB, for SAR, and between -0.1 and 0.1 with a step of 0.005, for the NDVI. For

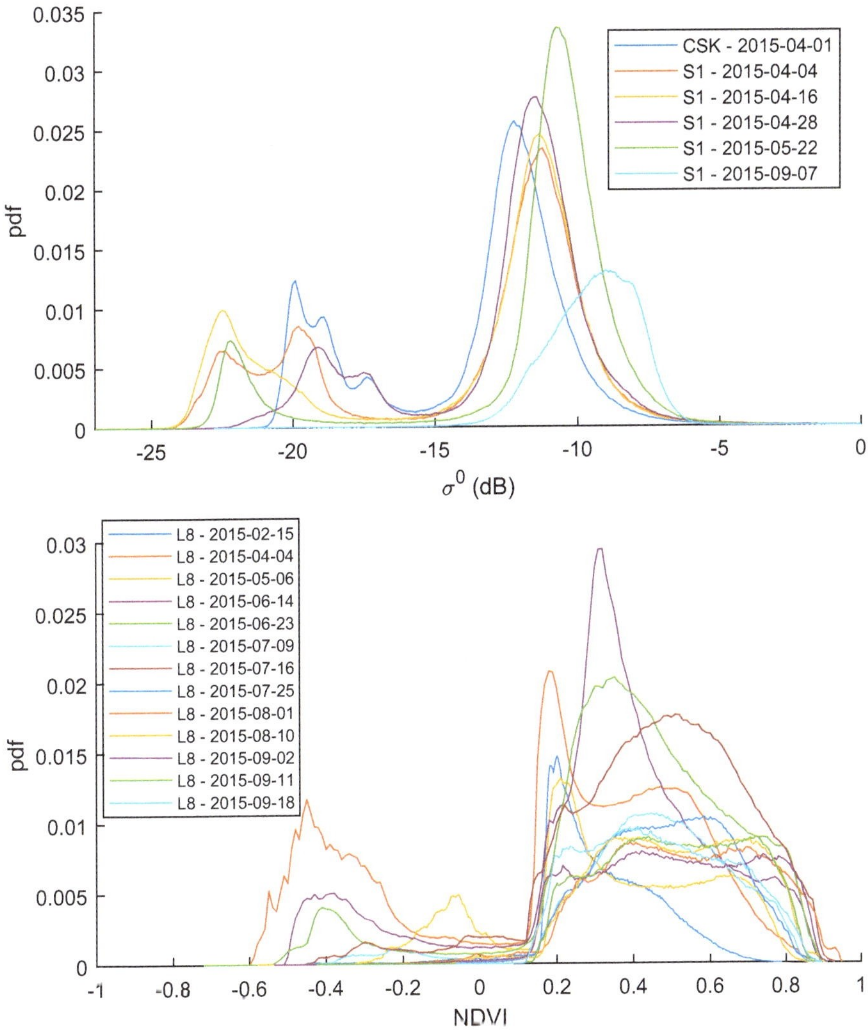

Fig. 4 Histograms of SAR backscatter coefficients (σ^0 from both COSMO-SkyMed and Sentinel-1) and Landsat 8 NDVI

each pair of thresholds, we compare the SAR and the optical binary flood map and compute a merit figure, given by Cohen's K value [4], to determine their agreement. As shown in Fig. 6, the K values calculated over the grid of pairs of SAR and NDVI thresholds form a smooth surface, with a single, broad maximum, located around the value pair [-15.7 dB (σ^0), 0.06 (NDVI)]. Given the relative smoothness of the K surface (see Fig. 6), we do not seek for finer positioning of the maximum and use the abovementioned pair of thresholds corresponding to the maximum K value to derive binary maps for each date. We use the derived σ^0 and NDVI optimal thresholds over

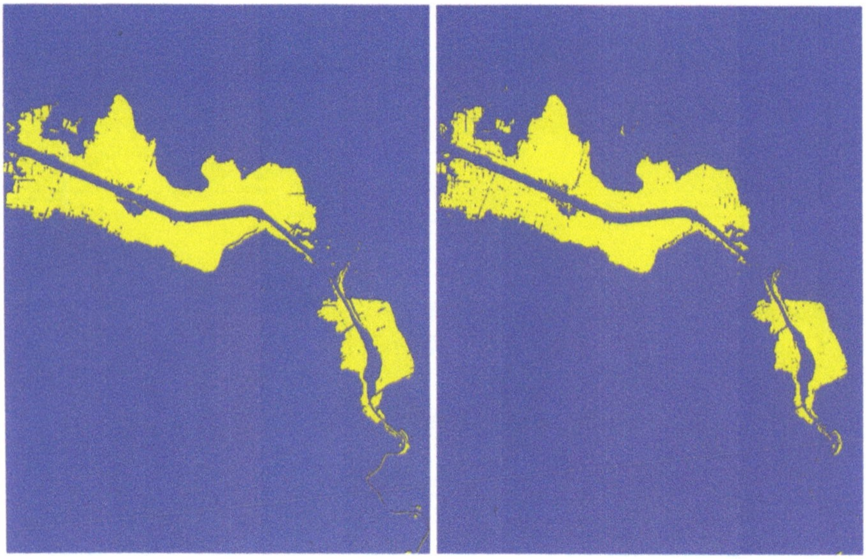

Fig. 5 (*Left*) Landsat 8 and (*right*) Sentinel-1 thresholded maps for the common date of April 4, 2015

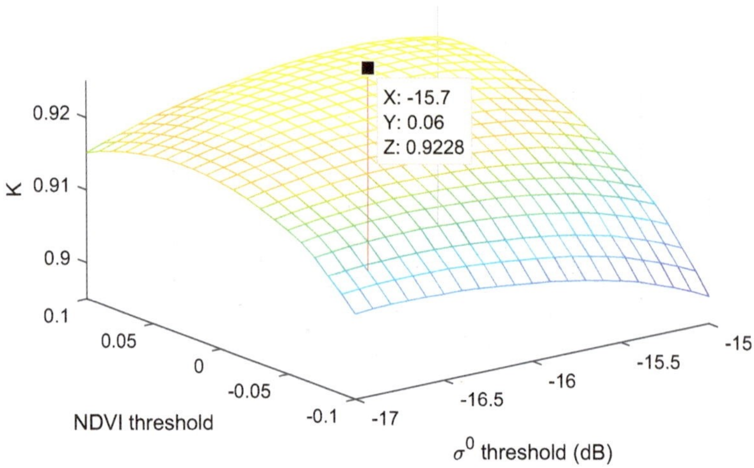

Fig. 6 Illustration of the method used to choose the best thresholds for the SAR backscatter and optical NDVI functions. The labeled point is the maximum of the surface, with the shown coordinates corresponding to the σ^0 threshold (X), the NDVI threshold (Y), and max(K) (Z)

all the SAR and optical images, respectively, including both the maps for the date of April 4, 2015, each with its optimal value, thus fully exploiting their slightly different acquisition times.

3 Results

The images help to reconstruct the post event flood recovery. The inundation affects the lowland areas, depressed with respect to the main channel. The areas remained flooded for long time, due to this particular geomorphological characteristic of this part of the river basin. The extreme rainfall occurred in the Bulgarian part of the basin, and subsequently the flood wave interested the depressed floodplain areas, which were inundated although the rainfall in the Greek part of the basin was not particularly intense.

The sequence of binary images constitutes a rather peculiar database. The time evolution of the flooded area can be followed throughout the image sequence. Figure 7 shows a combination of all the binary maps. Such combined map is obtained by simply summing up the 1/0 binary images of each date, thus assigning increasing indices to pixels being inundated in increasing numbers of dates with respect to the total time series. It is possible to note how peripheral areas are reached by the flood wave only on a few days around the maximum, while some pixels, close to the river course, stay flooded for most of the investigated dates, forming probably ponds which take long times to absorb and/or evaporate.

The total area occupied by flooded pixels in the region of interest can be computed and plotted in a graph, showing its evolution in time. The graph is shown

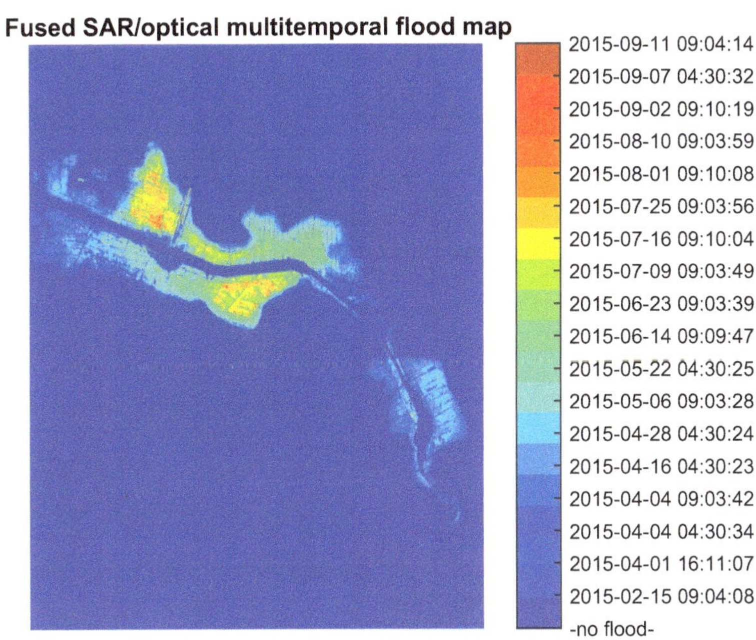

Fig. 7 Multisensor, multi-temporal flood extent map. Areas with different colors are flooded until the corresponding date on the *colorbar*

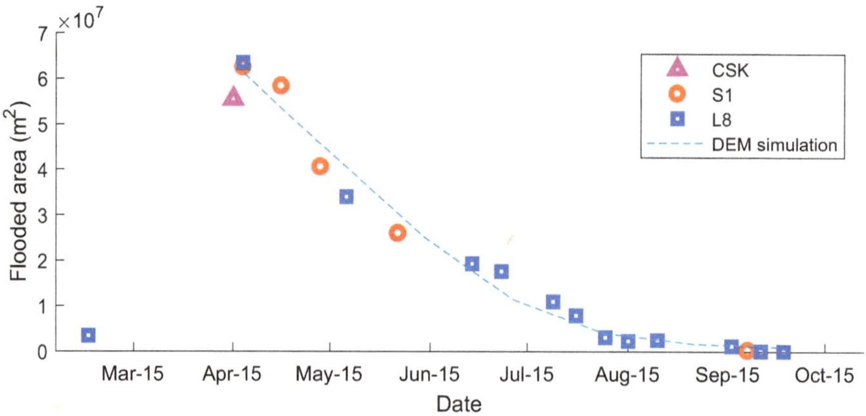

Fig. 8 Temporal trend of the flooded area. Symbols stand for information from different sensors. *CSK* COSMO-SkyMed, *S1* Sentinel-1; *L8* Landsat 8. *Dashed line* shows simulated data from the DEM (© JAXA)

in Fig. 8. The trend of this plot shows a relatively sharp increase in total area, from around 0 in February up to a maximum area of more than 6500 ha. Reported maximal flooded area by the Copernicus site (www.copernicus.eu) is 6576 ha, in very good agreement with the one estimated here. Data of this kind offer thus a rather novel tool to investigate phenomena such as inundation dynamics and evapotranspiration/infiltration processes, at an unprecedented spatial and temporal resolution.

For instance, we used data from the © JAXA Global World DEM, obtained from ALOS PRISM data [16, 17], to attempt understanding and reproducing the flood event dynamics. Figure 9 shows the DEM height maps. Missing data are visible in the upper part of the area, but the missing locations do not contribute to the flood area computation. Note, instead, the increased height recorded along the river course, which denotes the presence of elevated river levees. This is confirmed by close inspection of optical images. It can be noticed how the lowland area in the DEM map appears similar in shape to the maximum inundated area as visible from the SAR S1 and optical L8 images of April 4, 2015 (see Fig. 2b, h). We could then hypothesize that the flood phenomenon was controlled by the particular geomorphological conditions that determined the long-lasting presence of water on the floodplain. This seems justified by computing the ROC curve obtained by varying the height threshold and using the maximum flood extent binary map obtained from the L8 NDVI map in Fig. 2h as score (see Fig. 10). The area under the ROC curve is 0.92, thus suggesting a good prediction capability of the simple height threshold for the flood level.

To check the consistence of the mentioned conceptual process model for the flood event, we determined first the most likely height level reached by water, by computing, from the ROC curve, the optimal threshold, defined as that minimizing

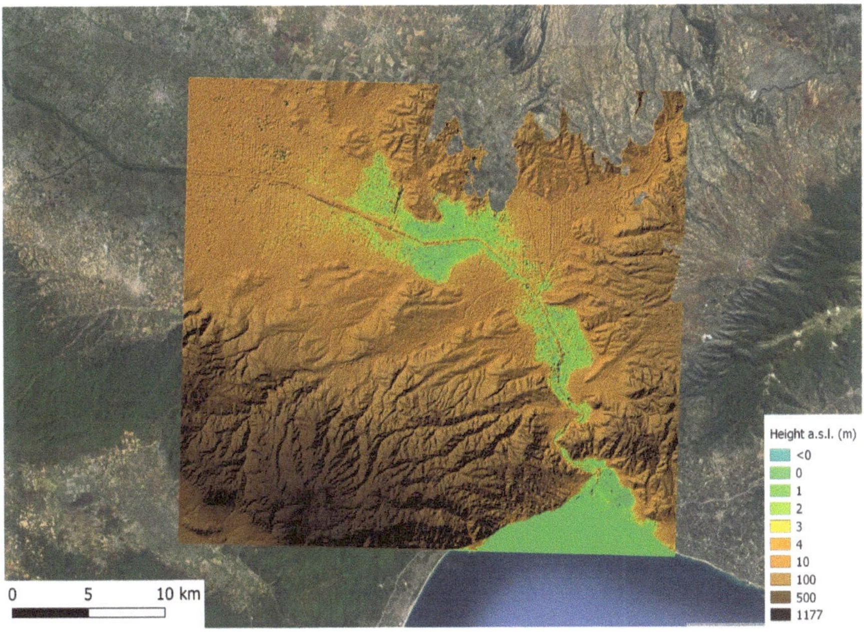

Fig. 9 ASTER DEM height map in hill-shaded view (© JAXA), draped on an optical image (© Google)

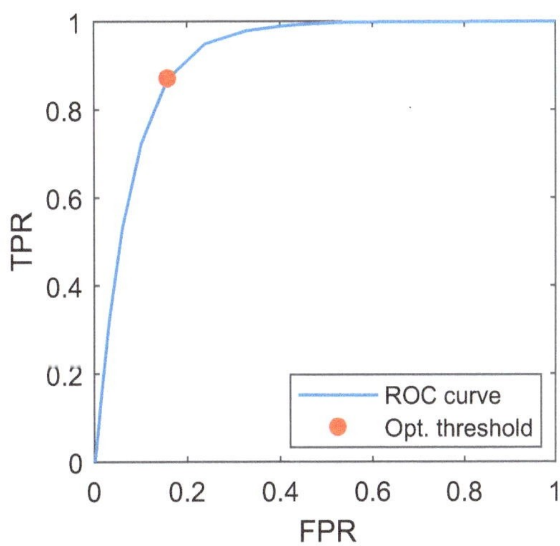

Fig. 10 ROC curve and optimal threshold value for the JAXA DEM threshold indicator with respect to the binary flood map obtained by the L8 NDVI image acquired on April 4, 2015. *FPR* false positive rate, *TPR* true positive rate. Area under the curve (AUC) is equal to 0.92

the misclassification cost [10], which results to be ∼4 m. Note that the DEM data is recorded in integer values, so no better accuracy is allowed in the threshold definition. We then simulated a progressive, steady decrease of the water level, linearly in time from the date of the maximum extent (April 4) to the end of the monitoring (September 2015), computing the total extents of the DEM area below each decreasing threshold. The resulting curve is plotted as a dashed line in Fig. 8. As can be seen, the overall agreement of the two trends is rather good, despite actual remotely sensed data show several oscillations with respect to the model. This may depend on several factors, e.g., the possible occurrence of rain throughout the monitored period – rain data are not available to us at the time of this writing – or water pumping in selected areas.

Acknowledgements The work is partly supported by the Apulia Space Project (PON-REC 2007–2013, C.N. PON03PE_00067_6). Sentinel-1 imagery is provided by the European Space Agency (ESA). COSMO-SkyMed imagery is provided by the Italian Space Agency (ASI) in the framework of Project (ID 2888/5229) "Spectral and temporal coherence for vessel detection and flood monitoring." DEM data © Japan Aerospace Exploration Agency (JAXA).

References

1. Alsdorf, D.E., Rodríguez, E., Lettenmaier, D.P.: Measuring surface water from space. Rev. Geophys. **45**(2), RG2002 (2007). doi:10.1029/2006RG000197
2. Bovolo, F., Bruzzone, L.: A split-based approach to unsupervised change detection in large-size multitemporal images: application to tsunami-damage assessment. IEEE Trans. Geosci. Remote Sens. **45**(6), 1658–1669 (2007). doi:10.1109/TGRS.2007.895835
3. Chander, G., Markham, B.: Revised landsat-5 tm radiometric calibration procedures and postcalibration dynamic ranges. IEEE Trans. Geosci. Remote Sens. **41**(11), 2674–2677 (2003). doi:10.1109/TGRS.2003.818464
4. Cohen, J.: A coefficient of agreement for nominal scales. Educ. Psychol. Meas. **20**(1), 37–46 (1960). doi:10.1177/001316446002000104
5. D'Addabbo, A., Refice, A., Pasquariello, G., Lovergine, F.P., Capolongo, D., Manfreda, S.: A Bayesian network for flood detection combining SAR imagery and ancillary data. IEEE Trans. Geosci. Remote Sens. **54**(6), 3612–3625 (2016). doi:10.1109/TGRS.2016.2520487
6. Deledalle, C.A., Denis, L., Tupin, F., Reigber, A., Jager, M.: NL-SAR: a unified nonlocal framework for resolution-preserving (Pol)(In)SAR denoising. IEEE Trans. Geosci. Remote Sens. **53**(4), 2021–2038 (2015). doi:10.1109/TGRS.2014.2352555
7. Faruolo, M., Coviello, I., Lacava, T., Pergola, N., Tramutoli, V.: A multi-sensor exportable approach for automatic flooded areas detection and monitoring by a composite satellite constellation. IEEE Trans. Geosci. Remote Sens. **51**(4), 2136–2149 (2013). doi:10.1109/TGRS.2012.2236336
8. Giustarini, L., Chini, M., Hostache, R., Pappenberger, F., Matgen, P.: Flood hazard mapping combining hydrodynamic modeling and multi annual remote sensing data. Remote Sens. **7**(10), 14200–14226 (2015). doi:10.3390/rs71014200
9. Grimaldi, S., Li, Y., Pauwels, V.R.N., Walker, J.P.: Remote sensing-derived water extent and level to constrain hydraulic flood forecasting models: opportunities and challenges. Surv. Geophys. **37**(5), 977–1034 (2016). doi:10.1007/s10712-016-9378-y
10. Hanley, J.A., Mcneil, B.J.: The meaning and use of the area under a receiver operating characteristic (ROC) curve. Radiology **143**(1), 29–36 (1982)
11. Malahlela, O.E.: Inland waterbody mapping: towards improving discrimination and extraction of inland surface water features. Int. J. Remote Sens. **37**(19), 4574–4589 (2016). doi:10.1080/01431161.2016.1217441
12. Martinis, S., Twele, A., Voigt, S.: Towards operational near real-time flood detection using a split-based automatic thresholding procedure on high resolution TerraSAR-X data. Nat. Hazards Earth Syst. Sci. **9**(2), 303–314 (2009). doi:10.5194/nhess-9-303-2009

13. Pulvirenti, L., Pierdicca, N., Boni, G., Fiorini, M., Rudari, R.: Flood damage assessment through multitemporal COSMO-SkyMed data and hydrodynamic models: the Albania 2010 case study. IEEE J. Sel. Top. Appl. Earth Obs. Remote Sens. **7**(7), 2848–2855 (2014). doi:10.1109/JSTARS.2014.2328012
14. Refice, A., Capolongo, D., Pasquariello, G., D'Addabbo, A., Bovenga, F., Nutricato, R., Lovergine, F.P., Pietranera, L.: SAR and InSAR for flood monitoring: examples with COSMO-SkyMed data. IEEE J. Sel. Top. Appl. Earth Obs. Remote Sens. **7**(7), 2711–2722 (2014). doi:10.1109/JSTARS.2014.2305165
15. Schreier, G. (ed.): SAR Geocoding: Data and Systems. Wichmann, Karlsruhe (1993)
16. Tadono, T., Nagai, H., Ishida, H., Oda, F., Naito, S., Minakawa, K., Iwamoto, H.: Generation of the 30 m-mesh global digital surface model by ALOS PRISM. ISPRS Int. Arch. Photogramm. Remote Sens. Spat. Inf. Sci. **XLI-B4**(July), 157–162 (2016). doi:10.5194/isprsarchives-XLI-B4-157-2016
17. Takaku, J., Tadono, T., Tsutsui, K., Ichikawa, M.: Validation of "AW3D" global DSM generated from ALOS PRISM. ISPRS Ann. Photogramm. Remote Sens. Spat. Inf. Sci. **III-4**(July), 25–31 (2016). doi:10.5194/isprsannals-III-4-25-2016
18. Twele, A., Cao, W., Plank, S., Martinis, S.: Sentinel-1-based flood mapping: a fully automated processing chain. Int. J. Remote Sens. **37**(13), 2990–3004 (2016). doi:10.1080/01431161.2016.1192304

Adaptive SAR Image Processing Techniques to Support Flood Monitoring from Earth Observation Data

Silvana G. Dellepiane and Laura Gemme

Abstract This chapter addresses the exploitation of Earth Observation (EO) data in the operational chains for flood monitoring and post-event damage assessment, focusing specifically on the task of post-flood mapping. In this context, this work provides a general review of our research into image processing techniques with an emphasis on adaptive methods applied to synthetic aperture radar (SAR) images. These procedures involve no restrictions on SAR acquisition parameters (frequency band, polarization, spatial resolution, and observation angle). Depending on the data availability, different maps can be produced. When multi-temporal images are available, two different products can be generated: *fast-ready flood maps;* and *detailed flood maps;*. The former is a color composite image that enhances the visualization of changes that have occurred after an event. The latter is a more detailed map obtained after a segmentation process. In contrast, when only an image acquired on a single date is available, a *water body map* can be generated. All these maps are intended as support for institutional interventions. Since only methods of segmentation and numerical data fusion are applied, such results are not final classification products. They are symbolic and not semantic maps, generated using fast and simple procedures that can be used as input for a classification purpose or employed by the user in other application tasks. The experiments described here were performed on real SAR images related to different datasets. The images were acquired from COSMO-SkyMed (CSK) and RADARSAT satellites.

Keywords SAR image segmentation • Adaptive image processing • Multi-temporal analysis • Data fusion • Color composite image • Fast-ready flood map

---

S.G. Dellepiane (✉) • L. Gemme
Department of Electrical, Electronic, Telecommunications Engineering and Naval Architecture (DITEN), Università degli Studi di Genova, Genova, Italy
e-mail: silvana.dellepiane@unige.it; laura.gemme@edu.unige.it

A. Refice et al. (eds.), *Flood Monitoring through Remote Sensing*,
Springer Remote Sensing/Photogrammetry,
https://doi.org/10.1007/978-3-319-63959-8_6

1 Introduction

Hydrological natural disasters such as floods are considered one of the most frequent calamities. Floods can have enormous human impact and can cause high levels of economic damage [1]. Different causes can contribute to flood events, including heavy rainfall, levee breaches, or dam failures. Therefore, during or after these events, a well-timed monitoring system is essential to help manage the situation.

In this framework, Earth Observation (EO) data acquired from space-based satellites can provide valuable and timely information that can be exploited to monitor and quantify the damage. Moreover, knowing flood extents is important for calibrating and validating hydraulic models [2]; detailed flood maps can also be assimilated into models.

Both optical and radar data can be employed as important tools for flood detection. The use of optical images is limited by weather conditions, in particular, by the presence of clouds. In contrast, synthetic aperture radar (SAR) images allow observation of wide areas with short revisit times and high resolution across all-weather and lighting conditions.

In general, flood detection algorithms aim to identify areas with low backscatter in SAR images. These regions behave as specular reflectors and appear as dark areas in a SAR image.

The authors of [3] include a discussion of backscattering responses and radar signatures of flooded terrain and observe that these common techniques may fail when vegetation protrudes from the water surface or when flooded urban areas must be detected. In these cases, both specular reflection and multiple reflections of the radar energy may occur. Furthermore, producing maps of bare terrain or completely inundated areas can become a challenging task. Finally, the presence of wind may increase the roughness of the water's surface, making it appear brighter and causing missed detections of flooded areas.

The main processing steps that may be used to generate flood maps from SAR images (not necessary in the following ordered sequence) are calibration, co-registration, feature extraction, segmentation, classification, and multi-temporal image analysis techniques.

The major goal of the work described in this chapter is to provide a general review of our recent research in image processing techniques devoted to the analysis of SAR intensity images in the context of flood monitoring [4] when neither calibration nor a model is used. The underlying basic concepts deal with the purpose of information extraction that requires minimal use of models, parameters, and threshold values with the objective of making automatic and optimizing the major image processing steps. To this end, methods that are unsupervised, adaptive, and independent from the order of analysis are proposed and applied, starting at the very beginning of the image processing chain and including histogram transformation, de-speckling filtering, segmentation, and multi-temporal image analysis.

These techniques can be applied to generate two product types [5]:

1. Fast-ready flood maps;
2. Detailed flood maps.

The first approach starts from a multi-temporal pair of images containing the same ground area. The proposed method generates these maps through color compositions that enhance the temporal evolution of inundated areas. These maps are intended as support for users who need to focus on rapidly acquiring the changes that occurred after the event.

In the second approach, a multi-temporal image segmentation is applied to generate detailed flood maps. Following the classical definition, image segmentation aims to partition an image into a set of different meaningful regions (perceptually homogeneous) based on certain homogeneity criteria. In this work, a bipartition of the image in "water" and "no-water" regions is considered. The proposed approach is based on the aggregation of pixels on the basis of the intensity similarity and growing the region of interest (ROI) through an adaptive mechanism that will be explained in greater detail in Sect. 3.2. The segmentation aims to provide more detailed information about water areas both before and after a flood event.

In contrast to the usual final products generated in this context [6] (e.g., the map of max. Water depth, the land cover map, the risk map, the vulnerability map, the damage map), these maps are not classification products; instead, they represent an intermediate added-value product intended to support authorities in their interventions or to be used as input for model-based systems.

The proposed procedures have no restrictions on acquisition mode or spatial resolution. In addition, because they employ few parameters, they prove to be simultaneously fast, accurate, and robust. The image content is aggregated by considering the contextual information, both spatial and temporal, whenever they exist.

By defining an appropriate test set, quantitative performance evaluations of flooded area detection can be performed; however, in this work, only qualitative evaluation is conducted, because a ground truth is not available.

This chapter is structured as follows. Section 2 addresses related works on flood monitoring from SAR images. Section 3 describes the proposed methodology, considering all the phases of the operational chain from the preprocessing steps to the flood maps. Section 4 presents the application of the proposed methods to real SAR images followed by a discussion of the results.

2 Related Works

Several methods dealing with flood monitoring have been published in the literature. These studies mainly focus on simple visual interpretation, histogram thresholding, image segmentation and classification, image texture approaches, and multi-temporal change detection methods (see, e.g., the references reported in [7]).

A simple visual interpretation consists of identifying changes between a pair of multi-temporal images through a simple color composition, allowing qualitative identification of flooded and non-flooded areas [8, 9].

Thresholding approaches [10–12], both manual and automatic, are used for fast visualization because they are computationally efficient. However, choosing the threshold is not simple due to the aforementioned backscattering signal considerations and because the choice of threshold depends on both environmental and system parameters (e.g., frequency band, polarization, and observation angle).

In addition to the thresholding methods, image segmentation has been considered to detect flooded areas and, subsequently, to delineate the boundaries of the flooded areas using methods such as active contours [13, 14]. Other segmentation approaches are semiautomatic and include the selection of a seed point [5, 15].

Kussul et al. [1] presented an approach for flood monitoring using SAR images from different satellites, applying artificial neural networks and Kohonen's self-organizing maps (SOMs) to perform image segmentation and classification.

Giordano et al. [16] proposed an innovative method using information provided by interferometric SAR images and applying a segmentation method based on fuzzy connectivity concepts in order to extract boundaries of the flooded area.

In [17], Pulvirenti et al. proposed an automatic segmentation method based on a morphological operator. The result was used to extract multi-temporal radar signatures from the identified regions, analyzing the radar backscattering using an electromagnetic model.

3 Methodology

As described earlier, the available EO data may include different contexts concerning flood monitoring. SAR image properties such as speckle noise, radar shadow, and multiple bounce effects can create unwanted artifacts and degrade the contrast, which makes image interpretation difficult. Therefore, data preparation through adaptive image processing techniques is necessary to improve the input for the subsequent steps, which include image segmentation and data fusion.

After data acquisition, operations such as calibration, geo-location, and orthorectification of SAR image data are typically performed to improve the overall map accuracy [18]. Moreover, speckle, which is intrinsically related to the backscattering signal, must be reduced using appropriate filters, and enhancing the data contrast can also improve the results of subsequent phases.

Depending on data availability, different products can be generated to detect flood areas and produce maps as support to authorities.

When both pre- and post-event images of the interested region are available, two processing products are proposed here that might either be used as flood maps or serve as the basis for flood map generation. These products are based on fusing the information content from both images. The first product (fast-ready flood map) is

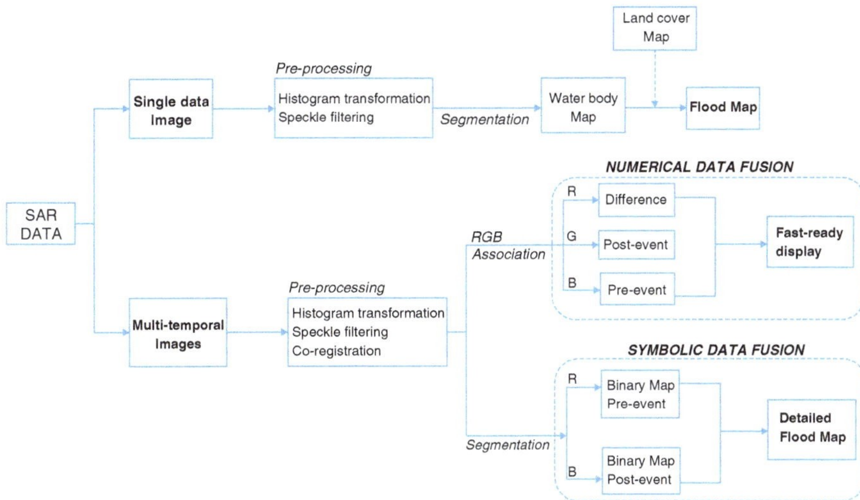

Fig. 1 The operation chain of the proposed methodology for flood monitoring

based on an RGB color composite, while the second product (detailed flood map) employs a segmentation algorithm.

When only one post-event image is available, it is possible to identify the water present in the image through segmentation techniques.

The operational chain for flood monitoring and post-event damage assessment includes the following technical steps:

- Histogram transformation
- Speckle filtering
- Graph-based segmentation
- Multi-temporal analysis

Figure 1 presents a block diagram of a general processing chain that extends from SAR data (input) to products (output). As shown, the process includes both a numerical and a symbolic data fusion.

3.1 Preprocessing

After image acquisition, some preprocessing steps, including calibration, co-registration, speckle filtering, and histogram transformation, are mandatory to provide improved input for the subsequent processing phases.

When multi-temporal images are available, spatial registration and precise calibration are usually required to obtain a useful image for displaying the changes that have occurred. Image calibration is a very complex and sensitive problem that involves two main phases, internal and external calibration, which include several operations. The entire procedure may be affected by measurement errors [19, 20]. In particular, errors can occur for the fluctuations in the transmitted power, receiver gains, system noise, and the illumination pattern of the antenna. To help solve the problem of performing a precise calibration step, we proposed a novel approach in [21] named "cross-calibration/normalization" that includes histogram truncation and equalization steps. The image histogram is clipped at gray levels higher than the 95th percentile, and then, an equalization step able to match the two histograms into a common uniform distribution model is performed. A generalization of this work was subsequently proposed by Nazir et al. [22] and Amitrano et al. [23].

Speckle is responsible for the granular aspect of coherent image systems and is caused by coherent processing of backscattered signals from multiple targets distributed within the resolution cell. Speckle is often considered a multiplicative noise, but it also contains significant information regarding the scene under observation. In the SAR community, various filters have been proposed for de-speckling purposes, such as those of Lee and Frost (and their enhanced versions), Kuan, Gamma MAP [24], and most recently, the speckle reducing anisotropic diffusion (SRAD) filter [25].

In the experiments, the SRAD filter has been applied because it preserves edges and information content. In addition, it overcomes some limitations of the classic filters, such as the sensitivity to the shape and size of the filter window and the capability to not enhance the edges. The SRAD filter is also adaptive and does not utilize thresholds. It depends on only two parameters: the number of iterations, n, and a "smoothing time step" parameter, λ, ranging between 0 and 1. We recently proposed a novel method to evaluate the performance of speckle filters, including some new statistical metrics and a new frequency-based approach. The results of this evaluation method showed that the SRAD filter produces the best results compared with several traditional filters [26].

3.2 *Single-Date Image Processing*

When only a single SAR image acquired on a post-flood date is available, a binary image product to identify water areas present in the scene can be generated.

This *water body map* can be used as a starting point for regional investigations. When a land cover map is available, it can be combined with the water body map to differentiate permanent water from flooded areas [4].

The presence of water can be localized through segmentation algorithms, which can be manual or automatic. In this work, we propose and briefly describe a semiautomatic approach. The segmentation method is an unsupervised process based on graph theory [27]. The majority of SAR segmentation techniques are manual or semiautomatic and supervised, with the necessity of a training set for different radar bands, image mode, and polarization [28]. In addition, they are based on statistical models and thresholding approaches [29], and they are affected by the estimation of parameters [30]. On the contrary, the proposed approach is adaptive to the image content and independent of models, a priori knowledge, and parameters. Therefore, it is both robust and repeatable. Furthermore, it does not involve restrictions on image modality acquisition, polarization, necessity of calibration, or spatial resolution.

The objective is to obtain a bipartition of the image into two subgraphs that are related to water and no-water areas, respectively. The method includes two phases: the cost-graph labeling phase and the graph-cut phase. The image is represented as an undirected, vertex-weighted grid graph $G(V, E)$, where V, the set of vertices, corresponds to the image pixels and the E set of edges represents links between neighboring pixels pairs. The algorithm starts from the insertion of a seed point s inside the region of interest, in this case water areas. The first step is to assign a weight to each vertex based on the absolute difference between the intensity of the seed point and a generic vertex. Then, the algorithm proceeds by finding optimal paths, named minimum path spanning tree (MPST), from the seed point to each vertex, assigning a cost value to each vertex. In particular, a spanning tree with respect to the following cost function [31]:

$$\mathrm{f}_{w_s}(v_i) = \min_{\pi(s,\mathrm{v_i})}\left[\max_{x\in\pi(s,v_i)} w_s(x)\right],$$

where $\pi\,(s, v_i)$ is the path from the source to a generic vertex v_i and w_s is the vertex weight, is generated.

Finally, to obtain a binary map, a cut of the graph is performed. By cutting the edges that show a cost larger than a given value (threshold), it is possible to obtain a subgraph $\mathrm{G}(V', E')$ representing a connected component corresponding to the ROI. The cost function and the criterion of graph visit guarantee that all vertices belonging to the ROI have a good path connecting them with the root in terms of intensity independent of spatial distance. When a tree shows a large difference between two consecutive levels, the paths have exceeded a local gradient. By changing the level of the graph cut, the tree can be interpreted as a multilevel segmentation. By choosing a low cut level, a small and very homogeneous region of interest can be obtained (since few cost values are taken into account); on the contrary, by choosing a high cut level, a large region of interest is extracted, potentially even covering the entire volume and with a more heterogeneous texture (since the ROI includes many cost values).

3.3 Multi-temporal Image Processing

The first product, the *fast-ready flood map*, consists of an RGB composition that helps enhance the changes in an image. It is automatically generated without user interaction, so it can support authorities during their initial interventions. The second product, the *detailed flood map*, is more accurate and involves two steps: a segmentation algorithm and an RGB association. Both these products can be considered sub-products of two more general mechanisms: a numerical data fusion and a symbolic data fusion, respectively. A brief description of both approaches is provided below.

The fast-ready flood map is a color composite image that provides a fast and reliable way to enhance changes that occurred after the event. Before color composition, cross-calibration/normalization based on histogram transformation (as in [21]) is applied. In contrast to other methods that use color composition [7], our approach is based on a systematic study of radar backscattering characteristics.

The color association is as follows:

- R: intensity image of difference between pre- and post-event images
- G: intensity post-event image
- B: intensity pre-event image

The image of difference is given by the following formula:

$$\mathrm{Diff}\,(i,j) = 128 + \frac{(i-j)}{2}$$

where i is the pixel value of the pre-event image and j is the pixel value of the post-event image. It follows from the definition that the image of difference has no negative values.

The composite image has the following characteristics [21]:

1. Low-backscattering areas from both images are displayed in dark red;
2. No-change, high-backscattering regions are rendered in cyan;
3. Results from decreased backscattering are rendered in magenta;
4. Changes due to an increase in backscattering are displayed in green.

Figure 2 shows a simple representation of this color association criterion, where the circle corresponds to the pre-event condition and the square is related to the post-event condition. Dark gray level means low backscattering, whereas light gray level means high backscattering. The intersections of these shapes represent the four possible cases previously described.

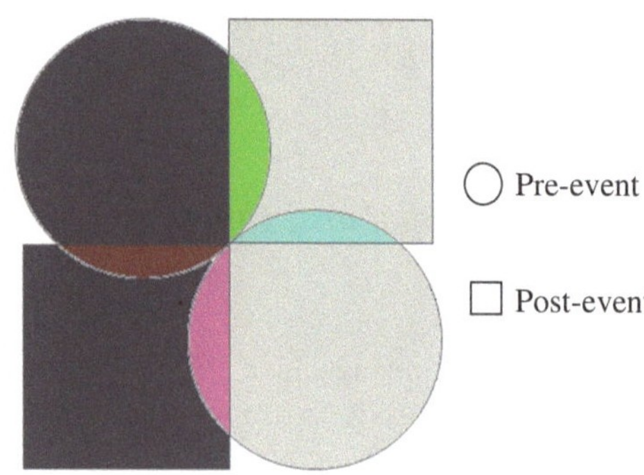

Fig. 2 Simple representation of the color fusion of the proposed method for the production of the fast-ready maps

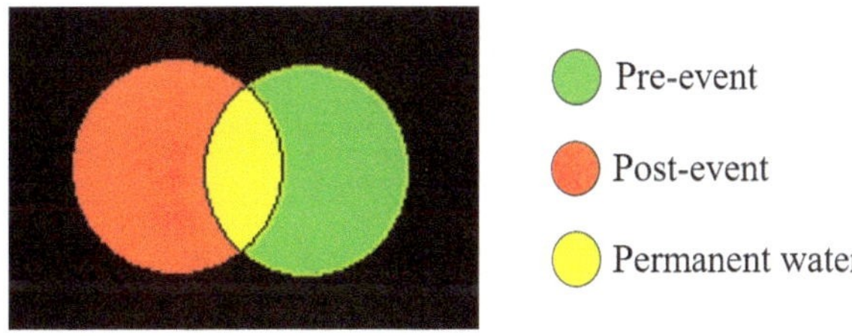

Fig. 3 Simple representation of color association for the generation of the detailed flood map

Detailed flood maps are generated after completing the image segmentation process described in Sect. 3.2. After applying the algorithm, a symbolic data fusion is carried out. The binary map of the pre-event image is associated with the green channel, whereas the binary map of the post-event image is associated with the red channel. The result is a color image in which yellow corresponds to permanent water (water areas identified in both images), red represents the low-backscattering areas present only after the event (floodwaters), and green identifies the low-backscattering areas present only before the event.

Figure 3 shows a schematic representation of the aforementioned color association.

4 Application

The experiments considered three case studies. Specifically, two flood events are considered. The former occurred in an area between Liguria and Tuscany (Italy), on October 2011. During that event, caused by a strong precipitation, about 542 mm of rain fell in just 6 h. Therefore, the Magra and the Vara rivers overflowed causing heavy damages in different towns. The latter took place in El Salvador in 2011, during the autumn months. Strong rains hit Central America forcing the authorities to declare a state of emergency and to ask for international aid. Many areas near the Lempa River were evacuated.

In addition, a dataset of multi-temporal images representing a rice paddy area that lies between Piedmont and Lombardy (Italy) is analyzed. The considered area is flat and rich of rivers and canals that allow abundant land irrigation, essential for the cultivation of rice. Because of their agricultural systems, rice fields can be used as a flood event simulation by considering "dry paddies" as similar to preflood conditions and "flooded paddies" as similar to post-flood conditions.

In this section, the considered datasets are first presented, and then, some experimental results are shown and discussed.

Table 1 Dataset of multi-temporal images of the flood event that occurred in Liguria, on October 25, 2011

Name	Acquisition date	Acquisition time	Look side	Orbit direction
CSK 1	13/5/2011	17:32:01	R	Descending
CSK 2	26/10/2011	17:13:03	R	Descending

Table 2 Dataset of images of the flood event that occurred in El Salvador

Name	Acquisition date	Acquisition time	Polarization	Orbit direction
RADARSAT 1	25/12/2001	11:46:57	HH	Descending
RADARSAT 2	22/10/2011	11:48:25	HH	Descending
CSK	22/10/2011	11:24:02	HH	Ascending

4.1 Datasets

Regarding the flood event that occurred on October 25, 2011, in Liguria, near the Magra River, a multi-temporal dataset consisting of two images is available (see Table 1): one before and one after the event. The images were acquired by the CSK constellation in StripMap HIMAGE mode and provided as geocoded terrain corrected, GTC_B level 1D products.

For the flood that occurred in El Salvador in October 2011, images acquired by different sensors were analyzed. A multi-temporal pair consisting of two RADARSAT images and a single-date image from the CSK mission (see Table 2) are available.

Concerning the multi-temporal pair, the available data were acquired on December 25, 2001, and October 22, 2011, respectively. Therefore, we made a comparison between a historical image and an image acquired during the period of strong rains, in order to highlight the changes. The image data are provided in two different product formats, SAR Georeferenced Fine (SGF) and SAR Systematic Geocoded (SSG), respectively. Both images are acquired in HH polarization.

The CSK image was acquired in StripMap HIMAGE modality on October 22, 2011, and provided as GTC_B product. The geometric resolution (both ground range and azimuth) is 5 m. The image polarization is HH.

The dataset representing rice fields consists of three pairs of images, each composed of a pre-event (2010) and one of three post-event (2011) acquisitions. The images were acquired from the CSK satellite constellation in StripMap HIMAGE mode (GTC_B product, level 1D). The geometric resolution (both ground range and azimuth) is 5 m. For the experiments, only the pair Pre_A – Post_3A has been employed, because the greater the interval between the times of image acquisition, the greater the differences that can be observed. For further information, Table 3 shows the main features of the whole dataset.

Table 3 Dataset of multi-temporal CSK images acquired in Piedmont and representing paddy fields

Name	Acquisition date	Beam off-nadir angle (°)	Polarization	Look side	Orbit direction
Pre_A	04/10/2010	24.13	HH	R	Ascending
Post_1A	27/05/2011	48.2	HH	R	Ascending
Post_2A	10/06/2011	38.56	HH	R	Ascending
Post_3A	16/06/2011	24.13	HH	R	Ascending

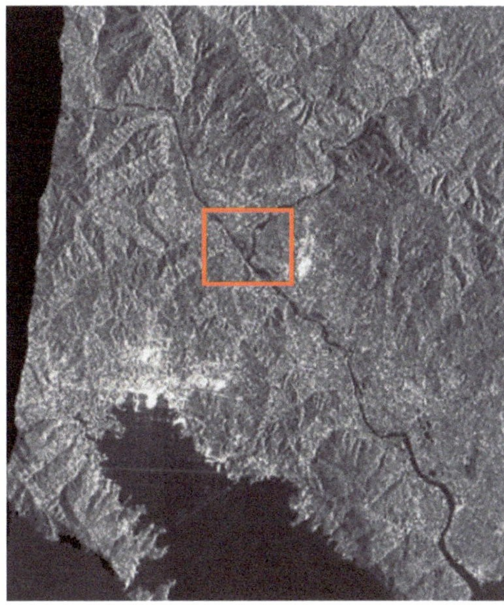

Fig. 4 The Liguria post-event image, acquired on October 26, 2011, highlighting the analyzed area of interest (*red rectangle*)

4.2 Experimental Results

This section presents and discusses some of the results of the operational chain, obtained after applying the proposed methods to the datasets introduced in the previous section.

4.2.1 Liguria Dataset

The first case study refers to the flood event occurred in Liguria on October 25, 2011. Figure 4 displays the SAR image acquired after the considered event, with the analyzed area evidenced by the red rectangle. Figure 5 shows the detailed view of the confluence of the Magra River with the Vara River (near La Spezia), acquired on May 13, 2011 (Fig. 5a), and on October 26, 2011 (Fig. 5b). By looking at the original images, it can be immediately noticed that the river is very swollen. The fast-ready map (Fig. 5c) helps noticeably in highlighting and revealing the flooded areas, mainly near the river (magenta color), compared to the other regions.

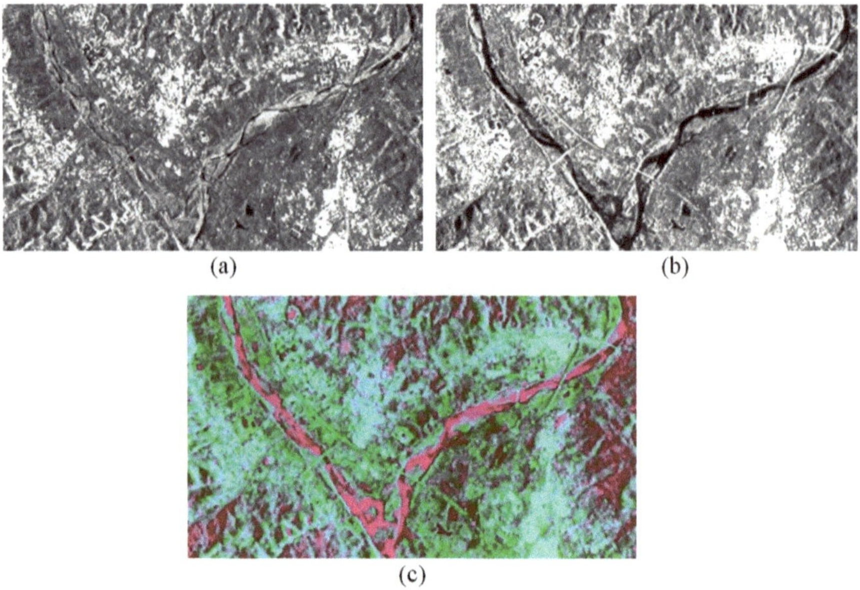

Fig. 5 Multi-temporal images of the flood that occurred near La Spezia (a) before and (b) after the flood event. (c) The corresponding fast-ready map

4.2.2 San Salvador Dataset

The second case study concerns the flood event that occurred in El Salvador. In October 2011, heavy rains from a tropical depression caused the break of banks in several points of Lempa River. One of the hardest hit areas includes the province of Usulután, in the southeast of the Country. In this section, we analyzed some of these zones close to the Lempa River, the Lake Suchitlán Cerrón Grande and the Jiquilisco Bay. More details about the areas affected by flooding can be found in the document [32].

The Lempa River overflowed in several places; therefore, we analyzed its path with greater attention.

Figure 6 shows the multi-temporal pair of images with three highlighted areas, that have been analyzed in detail (red, yellow and green rectangles).

Figure 7 displays the detail in the yellow rectangle, near the coast, whereas Fig. 8 shows the detail in the green rectangle. In both cases, in the fast-ready flood maps, it is evident the presence of several points where the river overflowed (in magenta color).

In addition to the areas along the path of the river previously analyzed, observing Fig. 6, it is evident that a change occurred also in the reservoir or Lake Suchitlán Cerrón Grande (see red rectangle). This is the largest lake in El Salvador, and it was formed with the creation of the Hydroelectric power plant of Cerrón Grande in the Lempa River. Figure 9 shows a detailed view over the lake Suchitlán Cerrón Grande

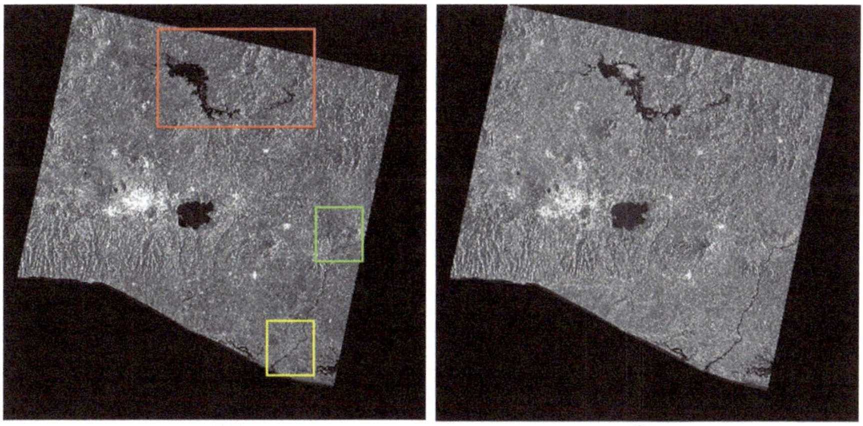

(a) (b)

Fig. 6 A multi-temporal pair of images acquired from the RADARSAT satellites, representing an area of El Salvador acquired on (**a**) December 2001 and on (**b**) October 2011

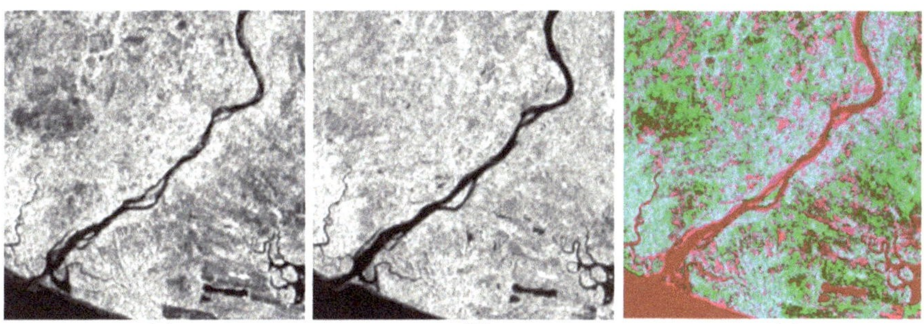

(a) (b) (c)

Fig. 7 A detail of the Lempa River, near the coast, (**a**) in a normal situation and (**b**) after strong rains. (**c**) The corresponding fast-ready map

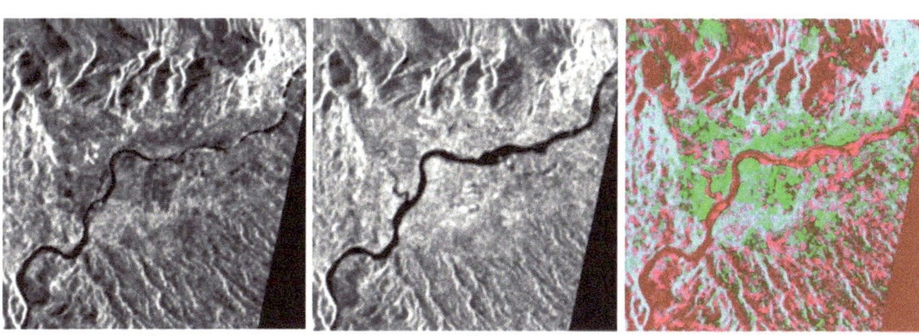

(a) (b) (c)

Fig. 8 A detail of the Lempa River, (**a**) in a normal situation and (**b**) after strong rains. (**c**) The corresponding fast-ready map

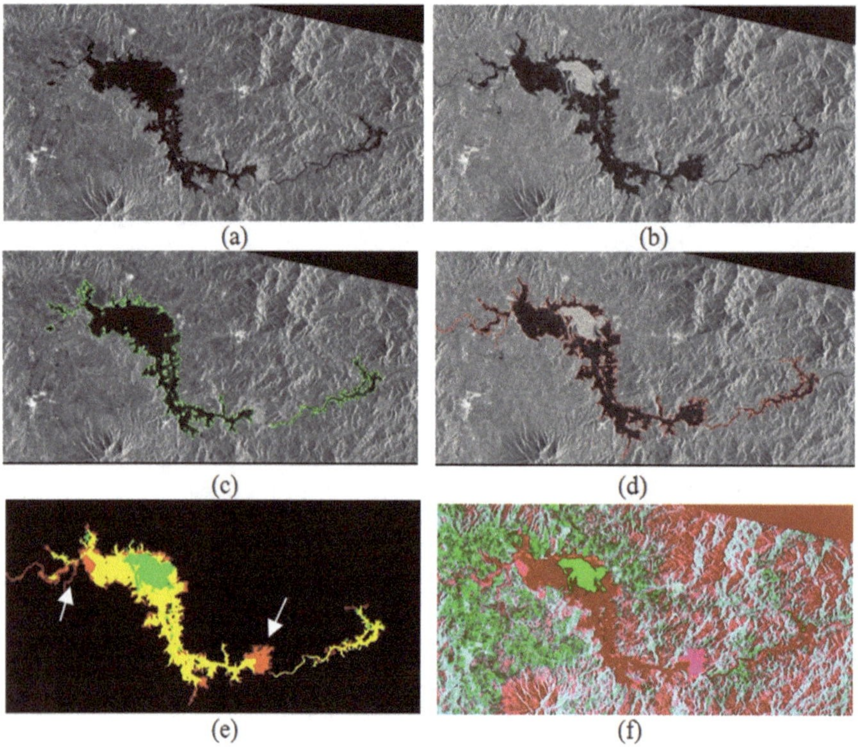

Fig. 9 Detailed views representing the lake Suchitlán Cerrón Grande (**a**) in a normal situation and (**b**) after a period of strong rains. The second row shows the (**c**) pre- and (**d**) post-event segmentation contours overlaid on the original images, and the third row shows the two proposed products: (**e**) the detailed map and (**f**) the fast-ready map

of the images shown in Fig. 6, acquired on December 2001 and October 2011, respectively (Fig. 9a, b). Because a multi-temporal pair of images was available, both products of the proposed method can be generated.

After the manual assignment of six seed points placed inside the water areas, the segmentation algorithm starts, and two binary images are obtained. For a better visualization, the contours of the segmented images are overlaid on the original image (Fig. 9c, d). Then, the two binary maps are associated to the green and red colors, generating the detailed flood map (Fig. 9e). The yellow regions correspond to water in both areas (permanent water). Areas in green color correspond to low backscattering in the pre-event image, whereas areas in red color correspond to low backscattering in the post-event image.

The fast-ready flood map (Fig. 9f) focuses attention on the extent of the flooded areas while enhancing the changes that have occurred (as well as indicating stable water). In particular, areas in green and magenta color indicate changes occurred after the event. The green color corresponds to areas where the backscattering signal is increased (probably due to the vegetation or water areas from the first date that were dry on the second date). The magenta color is related to areas where the backscattering is decreased (possible flooded zones).

Thanks to both products, it is possible to highlight the areas where changes occurred. The detailed map allows to better visualize these areas (as indicated by the white arrows in Fig. 9e, in correspondence of the dam in the lower part and of agricultural fields in the upper part).

Specifically, the fast-ready map is a raster map, whereas the detailed map is a vector map. In fact, the detailed map allows subsequent processing such as the contour identification, the generation of a vector map, and the computation of some measurements, for instance, related to the surface.

Finally, Fig. 10a shows the original CSK single-date image, acquired on October 22, 2011, at 11:24, with a focus on the Jiquilisco Bay (see blue rectangle). This area is located on the southeast Pacific coast, in the Department of Usulután, and is a reserve which includes the largest mangroves in El Salvador and several bays, canals, and forests. In addition, it has a relevant importance in the prevention of natural disasters, such as floods and earthquakes.

During the flooding event of October 2011, Jiquilisco was the municipality with the largest number of displaced persons, 30% of its total population [32].

Figures 10b, c show, respectively, a detailed view of the Jiquilisco Bay and the corresponding water body map, after applying the segmentation method described in Sect. 3.2. For the segmentation, eight seed points have been utilized to generate the cost image, and finally, a cut of the tree has provided the binary map.

In this case, the "water class" includes both permanent water (such as rivers) and changed areas (such as floodwaters). This preliminary result can later be integrated with a land cover map or with a priori information about the area under observation.

4.2.3 Piemonte Dataset

Figure 11 displays details of rice paddy fields near the Po River from the third dataset and some intermediate processing results. Figures 11a, b show the multi-temporal image pair acquired on October 4, 2010 (Pre_A of Table 3), and June 16, 2011 (Post_3A of Table 3), respectively. Figures 11c, d show preprocessed images after applying the histogram transformation and the SRAD filter with $n = 5$ and $\lambda = 0.8$. As can be observed, speckle is reduced and contrast is enhanced.

Finally, Figs. 11e, f show the generated fast-ready maps; the former was obtained applying the color association explained in Fig. 2 (Sect. 3.3) to the original images without any elaboration, whereas the latter is the result after applying the same color association to the processed images (filtering and histogram truncation). The dark red color represents permanent water such as rivers or lakes (weak scatter) on both dates. The cyan regions correspond to unchanged no-water areas such as cities (strong scatter). The magenta color indicates decreased backscattering, which can indicate where the water level has risen to cover the rice plants. The green

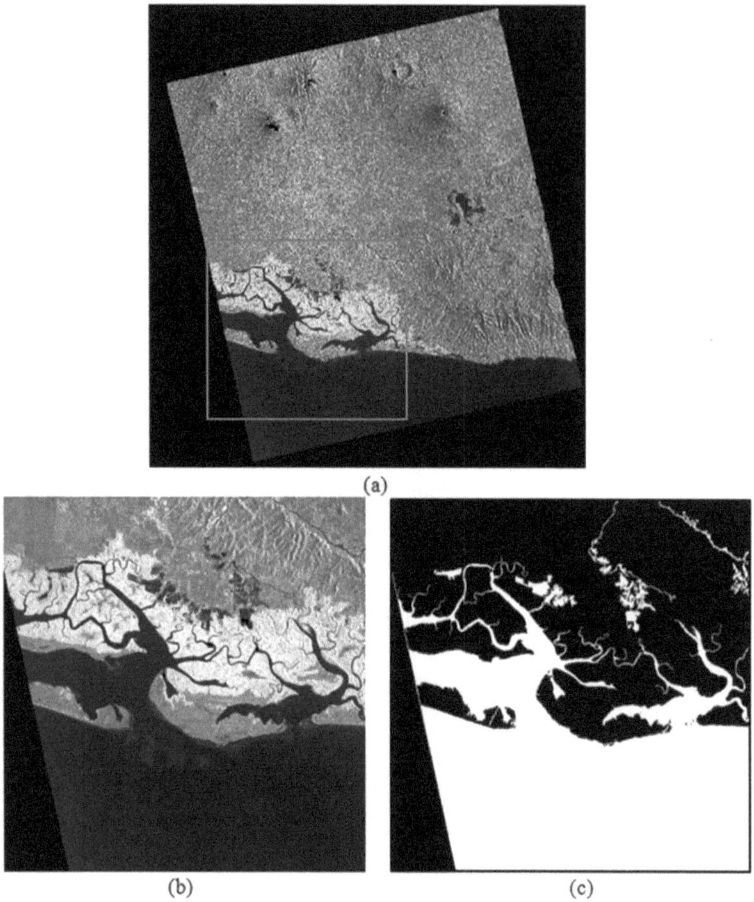

Fig. 10 (a) The original image CSK image acquired on October 22, 2011. (b) The detailed view of the Jiquilisco Bay and c the corresponding binary map

color represents increased backscattering, which can be related to paddy drying or to vegetation growth.

Furthermore, it can be observed that the image result obtained after applying the preprocessing phase (Fig. 11f) simplifies analyzing the small details and enhances the visual interpretation.

5 Conclusions

Flood monitoring and detection are important aspects of providing support for civil protection and other authorities. In this context, EO data can play an important role,

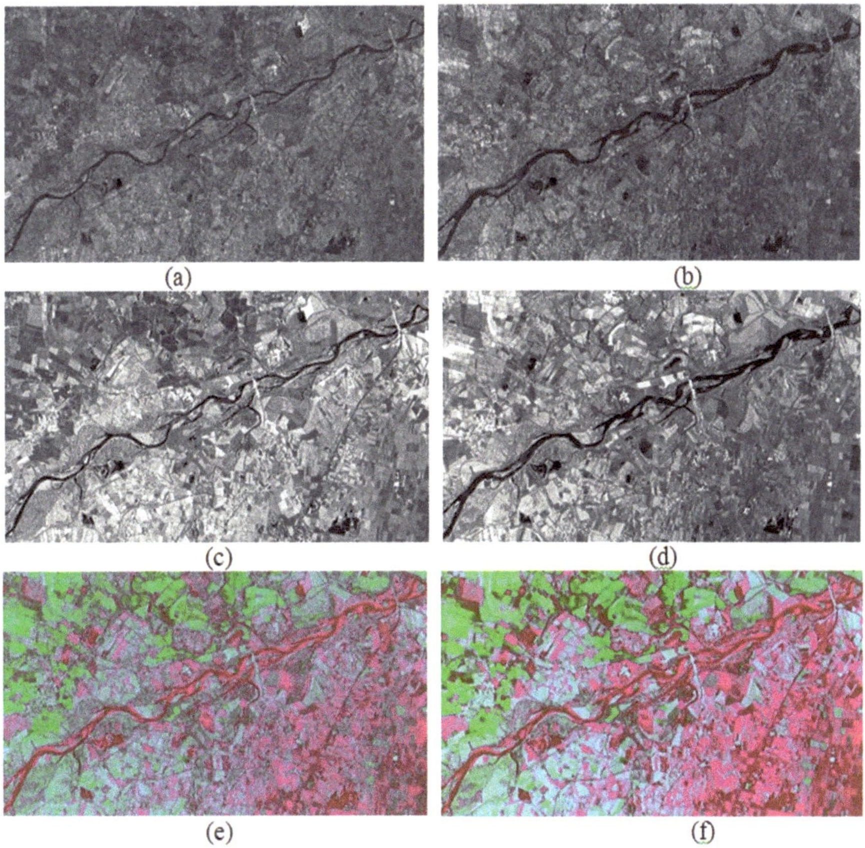

Fig. 11 Some experimental results from the dataset representing rice paddy fields. (**a**, **b**) Original CSK images acquired on October 4, 2010 (pre-event), and June 16, 2011 (post-event), respectively; (**c**, **d**) the previous images, after applying histogram transformation and de-speckle filtering; (**e**) fast-ready map starting from the original images, without any elaboration; (**f**) fast-ready map after applying the histogram transformation and the de-speckle filtering to the original images

especially the SAR images that allow continuous and effective observations of the Earth.

In this chapter, various image processing techniques have been proposed to support flood monitoring and damage assessment. These techniques can adapt to the image content and are independent of the image acquisition mode, spatial resolution, polarization, and other parameters. In particular, an operational chain targeted toward flood mapping is proposed. Depending on data availability, three possible products are suggested. These products are not classification maps, since we applied symbolic associations; they are intermediate products intended to provide benefit to the authorities to support interventions or to be used as input for subsequent model-based systems.

When only an image from a single date is available, a water body map is generated. In this case, the result is a bipartition of the image into water and no-water areas. It can be considered as a preliminary result that can be integrated with a land use map or with other information.

The second approach starts from a multi-temporal pair of images of the same ground area and results in a fast-ready flood map consisting of a color composite image that provides a fast and easy way to identify the changes that occurred after an event. The third approach, also based on multi-temporal data, is generated after an initial graph-based segmentation process and results in detailed flood maps. These are more precise maps that highlight both steady (permanent) water and changes. This last product can be used as an additional information layer in geographical information systems for event management. Furthermore, the detailed flood maps can be assimilated into complex hydrological prevision models to take advantage of the EO data.

Acknowledgments This work is framed in the context of the "OPERA—Civil protection from floods" pilot project funded by the Italian Space Agency in cooperation with the Italian Department for Civil Protection.

References

1. Kussul, N., Shelestov, A., Skakun, S.: Flood monitoring from SAR data. In: Use of Satellite and In-Situ Data to Improve Sustainability, pp. 19–29. Springer, Netherlands (2011)
2. Horritt, M.S.: A methodology for the validation of uncertain flood inundation models. J. Hydrol. **326**(1), 153–165 (2006)
3. Pierdicca, N., Pulvirenti, L., Chini, M., Boni, G., Squicciarino, G., Candela, L.: Flood mapping by SAR: possible approaches to mitigate errors due to ambiguous radar signatures. In: IEEE International Geoscience and Remote Sensing Symposium (IGARSS), Quebec City, Canada, pp. 3850–3853. IEEE (Institute of Electrical and Electronics Engineers Inc., Piscataway, New Jersey, US) (2014)
4. Dellepiane, S., Angiati, E.: COSMO-SkyMed in support of flood monitoring. In: IEEE International Geoscience and Remote Sensing Symposium (IGARSS), Munich, pp. 778–781. IEEE (Institute of Electrical and Electronics Engineers Inc.) (2012)
5. Dellepiane, S., Angiati, E., Vernazza, G.: Processing and segmentation of COSMO-SkyMed images for flood monitoring. In: IEEE International Geoscience and Remote Sensing Symposium (IGARSS), Hawai, US, pp. 4807–4810. IEEE (Institute of Electrical and Electronics Engineers Inc., Piscataway, New Jersey, US) (2010)
6. Boni, G., Candela, L., Castelli, F., Dellepiane, S., Palandri, M., Persi, D., . . . , Versace, C.: The OPERA project: EO-based flood risk management in Italy. In: IEEE International Geoscience and Remote Sensing Symposium (IGARSS), Cape Town, South Africa, pp. 929–932. IEEE (Institute of Electrical and Electronics Engineers Inc., Piscataway, New Jersey, US) (2009)
7. Schumann, G., Bates, P.D., Horritt, M.S., Matgen, P., Pappenberger, F.: Progress in integration of remote sensing–derived flood extent and stage data and hydraulic models. Rev. Geophys. **47**(4), RG4001 (2009)
8. Martinez, J.M., Le Toan, T.: Mapping of flood dynamics and spatial distribution of vegetation in the Amazon floodplain using multitemporal SAR data. Remote Sens. Environ. **108**(3), 209–223 (2007)

9. Dellepiane, S. G., Bo, G., Monni, S., Buck, C.: Improvements in flood monitoring by means of interferometric coherence. In: SAR Image Analysis, Modeling, and Techniques III, Barcelona, Spain, pp. 219–229. Proceedings of SPIE - The International Society for Optical Engineering (Bellingham, Washington, US) (2000)
10. Martinis, S., Twele, A., Voigt, S.: Towards operational near real-time flood detection using a split-based automatic thresholding procedure on high resolution TerraSAR-X data. Nat. Hazards Earth Syst. Sci. **9**(2), 303–314 (2009)
11. Matgen, P., Hostache, R., Schumann, G., Pfister, L., Hoffmann, L., Savenije, H.H.G.: Towards an automated SAR-based flood monitoring system: lessons learned from two case studies. Phys. Chem. Earth A/B/C. **36**(7), 241–252 (2011)
12. Long, S., Fatoyinbo, T.E., Policelli, F.: Flood extent mapping for Namibia using change detection and thresholding with SAR. Environ. Res. Lett. **9**(3), 035002 (2014)
13. Horritt, M.S., Mason, D.C., Luckman, A.J.: Flood boundary delineation from synthetic aperture radar imagery using a statistical active contour model. Int. J. Remote Sens. **22**(13), 2489–2507 (2001)
14. Mason, D.C., Horritt, M.S., Dall'Amico, J.T., Scott, T.R., Bates, P.D.: Improving river flood extent delineation from synthetic aperture radar using airborne laser altimetry. IEEE Trans. Geosci. Remote. Sens. **45**, 3932–3943 (2007)
15. Martinis, S., Twele, A., Voigt, S.: Unsupervised extraction of flood-induced backscatter changes in SAR data using Markov image modeling on irregular graphs. IEEE Trans. Geosci. Remote Sens. **49**(1), 251–263 (2011)
16. Giordano, F., Goccia, M., Dellepiane, S.: Segmentation of coherence maps for flood damage assessment. In: IEEE International Conference on Image Processing 2005, ICIP 2005, Genova, Italy, vol. 2, pp. 233–236. IEEE (Institute of Electrical and Electronics Engineers Inc., Piscataway, New Jersey, US) (2005)
17. Pulvirenti, L., Chini, M., Pierdicca, N., Guerriero, L., Ferrazzoli, P.: Flood monitoring using multi-temporal COSMO-SkyMed data: image segmentation and signature interpretation. Remote Sens. Environ. **115**(4), 990–1002 (2011)
18. Angiati, E., Dellepiane, S., De Martino, M., Moser, G., Serpico, S.B.: Flooding and change maps from COSMO-SKYMED images, Riunione Annuale GTTI. www.gtti.it/gtti10/papers/gtti10_submission_45.pdf (2010)
19. Freeman, A.: SAR calibration: an overview. IEEE Trans. Geosci. Remote Sens. **30**(6), 1107–1121 (1992)
20. Shimada, M.: Verification processor for SAR calibration and interferometry. Adv. Space Res. **23**(8), 1477–1486 (1999)
21. Dellepiane, S.G., Angiati, E.: A new method for cross-normalization and multitemporal visualization of SAR images for the detection of flooded areas. IEEE Trans. Geosci. Remote Sens. **50**(7), 2765–2779 (2012)
22. Nazir, F., Riaz, M.M., Ghafoor, A., Arif, F.: Brief communication: contrast-stretching-and histogram-smoothness-based synthetic aperture radar image enhancement for flood map generation. Nat. Hazards Earth Syst. Sci. **15**(2), 273–276 (2015)
23. Amitrano, D., Di Martino, G., Iodice, A., Riccio, D., Ruello, G.: A new framework for SAR multitemporal data RGB representation: rationale and products. IEEE Trans. Geosci. Remote Sens. **53**(1), 117–133 (2015)
24. Shi, Z., Fung, K. B.: A comparison of digital speckle filters. In: Proceedings of IGARSS '94-1994 IEEE International Geoscience and Remote Sensing Symposium, vol. 4. IEEE; pp. 2129–2133. doi:10.1109/IGARSS.1994.399671 (1994)
25. Yu, Y., Acton, S.T.: Speckle reducing anisotropic diffusion. IEEE Trans. Image Process. **11**(11), 1260–1270 (2002)
26. Dellepiane, S.G., Angiati, E.: Quality assessment of despeckled SAR images. IEEE J. Sel. Top. Appl. Earth Obs. Remote. Sens. **7**(2), 691–707 (2014)
27. Peng, B., Zhang, L., Zhang, D.: A survey of graph theoretical approaches to image segmentation. Pattern Recogn. **46**(3), 1020–1038 (2013)

28. Zhang, B., Perrie, W., Li, X., Pichel, W.G.: Mapping sea surface oil slicks using RADARSAT-2 quad-polarization SAR image. Geophys. Res. Lett. **38**(10), (2011)
29. Feng, J., Cao, Z., Pi, Y.: Multiphase SAR image segmentation with-statistical-model-based active contours. IEEE Trans. Geosci. Remote Sens. **51**(7), 4190–4199 (2013)
30. Derrode, S., Mercier, G.: Unsupervised multiscale oil slick segmentation from SAR images using a vector HMC model. Pattern Recogn. **40**(3), 1135–1147 (2007)
31. Gemme, L., Dellepiane, S.: A new graph-based method for automatic segmentation. Lecture Notes in Computer Science (including subseries Lecture Notes in Artificial Intelligence and Lecture Notes in Bioinformatics), 9279, 18th International Conference on Image Analysis and Processing, ICIAP 2015, Genova, Italy, pp. 601–611. Springer Verlag, Cham, Switzerland (2015)
32. Emergency appeal n°MDRSV004, El Salvador: Floods. November 4, 2011. Available: reliefweb.int/sites/reliefweb.int/files/resources/MDRSV004ea.pdf

Flood Mapping in Vegetated and Urban Areas and Other Challenges: Models and Methods

Nazzareno Pierdicca, Luca Pulvirenti, and Marco Chini

Abstract Floods are the most frequent weather disasters in the world and the most costly in terms of economic losses. Mapping flood extension is fundamental to ascertain the damage and for relief organization. Spaceborne synthetic aperture radar (SAR) systems represent a powerful tool to monitor floods because of their all-weather capability, the very high spatial resolution of the new generation of instruments, and the short revisit time of the present and future satellite constellations. However, mapping flooded vegetated and urban areas still represents a challenging problem.

Modeling different targets both in the presence and in the absence of flood water is a very complex task. In the first part of the chapter we review these challenging conditions, showing their potential effects on radar data and in particular on COSMO-SkyMed images. In some cases the potential of electromagnetic models to predict the radar response is shown.

A second part of the chapter illustrates a number of strategies one can exploit, with examples showing the achievable performances with respect to a simple mapping of dark areas in the SAR image.

Keywords SAR backscatter modeling • SAR-based flood monitoring • SAR polarimetry and interferometry for flood mapping

N. Pierdicca (✉)
Department of Information Engineering, Electronics and Telecommunications, Sapienza University of Rome, Rome, Italy
e-mail: nazzareno.pierdicca@uniroma1.it

L. Pulvirenti
CIMA Research Foundation, Savona, Italy

M. Chini
Luxembourg Institute of Science and Technology, Belvaux, Luxembourg

A. Refice et al. (eds.), *Flood Monitoring through Remote Sensing*,
Springer Remote Sensing/Photogrammetry,
https://doi.org/10.1007/978-3-319-63959-8_7

1 Introduction

Floods are the most frequent weather disasters in the world and the most costly in terms of economic losses. Mapping flood extension is fundamental to ascertain the damage and for relief organization. Spaceborne synthetic aperture radar (SAR) systems represent a powerful tool to monitor floods because of their all-weather capability, the very high spatial resolution of the new generation of instruments, and the short revisit time of the present and future satellite constellations.

The main mechanism producing a high contrast in the radar image between flooded and non-flooded terrain is the specular reflection from standing water leading to a very low radar backscatter. However, the radar signatures of the Earth surface are often ambiguous, depending on several factors (e.g., structure and geometry of the target, presence, and type of vegetation) that are generally unknown. For instance, mapping flooded vegetated and urban areas represents a challenging problem. Other challenges that may be encountered are related to surface topography (e.g., false alarms due to shadows), to particular land covers (e.g., wet snow, very dry soils), and to atmospheric conditions (heavy precipitation when radar frequency is high, or high wind). In this chapter, we review these challenging conditions and explain in an intuitive manner the scattering mechanisms that affect the radar response. The literature about electromagnetic models able to predict quantitatively the microwave response of microwave sensors is reviewed. In addition, we present some methods developed to cope with these problems as well as case studies where these methods were successfully exploited.

2 Radar Response of Flooded Land

2.1 *Radar Signature of Flooded Bare Soil*

The capability to detect and monitor floods using microwave sensors originates from the very high sensitivity of the microwave interaction with natural media to the presence of water in the media itself. Liquid water is a highly dissipative medium with large values of dielectric constant. The latter is determined by the combined effect of dielectric relaxation of water molecular dipoles and ohmic losses in the presence of electric conductivity. In fact, water modifies the dielectric properties of the earth surface (e.g., it increases the permittivity of the soil) but also its geometric characteristics (e.g., the roughness of the surface in the presence of standing water as compared to that of soil).

The major mechanism producing a high contrast in the radar image between flooded and non-flooded terrain is the predominant specular reflection from the smooth water surface having high permittivity, as compared to the diffuse scattering from a rough and dry soil, as depicted in Fig. 1. Specifically, the smooth water surface covering the terrain reflects the impinging radar signal mainly in the specular

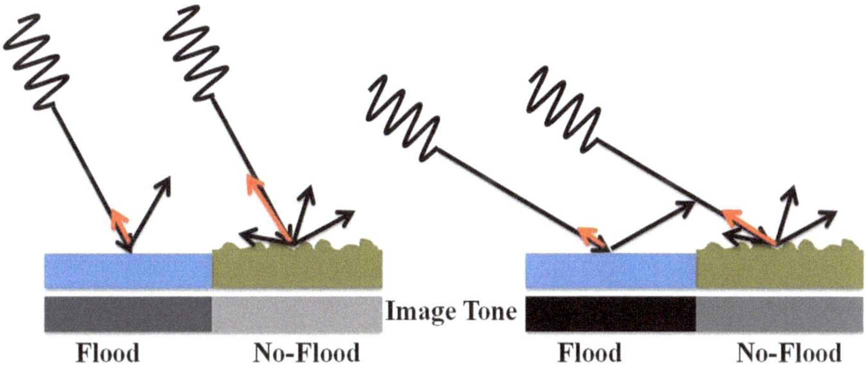

Fig. 1 Scattering mechanisms determining the radar signature of a flooded bare terrain for low (*left*) and high (*right*) incidence angle

direction (i.e., the direction forming an angle with respect to the perpendicular to the surface equal to that of the incidence wave). This produces a sharp scattering pattern around this direction and a very low backscatter measured by a monostatic radar, which implies a dark tone in the radar image. Conversely, the rough non-flooded terrain scatters the signal in many different directions and may produce a higher scattering coefficient (σ^0) in the backward direction too, which determines a brighter image tone. This mechanism is therefore driven mainly by the difference in roughness between standing water and terrain. However, the different permittivity plays also a significant role. In simple terms, the higher permittivity of the standing water increases the contrast between the electromagnetic wave impedance of the upper media (air) and that of the terrain. This implies that the total fraction of reflected power (i.e., the so-called surface albedo) increases with respect to that of dry soil. This effect has a strong impact on the surface emissivity, i.e., one minus the surface albedo, according to the Kirchhoff principle. The presence of water increases the surface albedo and thus significantly decreases the emissivity and the natural thermal emission detected, e.g., by a microwave radiometer (i.e., the so-called brightness temperature). In a radar image, although the albedo of the water surface is larger, the scattered energy is predominantly directed in the specular direction, and then the image taken by a monostatic radar appears dark. Note that the situation may change with a bistatic radar, if the scattered signal is observed in a direction very different from the one of the illuminating antenna. Moreover, when illuminating a rough surface, the amount of electromagnetic energy not reflected back in the upper hemisphere goes across the surface boundary. It is of minor interest for the imaging radar, as it is generally dissipated into the soil, unless there are inhomogeneities or stratifications that produce some volume scattering.

In practice, there is always a mixture of coherent reflection in the specular direction and incoherent diffuse scattering in any direction, as even standing water is not perfectly smooth. Considering the irradiation pattern from a rough surface, including coherent specular and diffuse incoherent components, it can be demonstrated that the contrast between dark flooded areas and bright rough bare soil is generally higher for larger incidence angles, as schematically depicted in Fig. 1 [62]. Put simply, the scattering from a smooth surface decreases more sharply as a function of the incidence angle than that of a rough surface. This is due to the prevalence of the coherent reflection, which has a very sharp angular pattern, although not exactly unidirectional. Then, the water surface is brighter when observed at steep angles and thus less contrasted with respect to the terrain characterized by a wider scattering pattern. However, in the works using current satellite radar systems, with intermediate values of the incidence angle, this effect was rarely noticed and becomes more complicated when the soil is covered by vegetation (see below).

It must be also considered that the volumetric soil moisture content (*SMC*) increases the complex soil permittivity [38, 40], and this implies an increase of the contrast between the electromagnetic impedances of air and terrain. Consequently, for a constant roughness, the backscattered σ^0 increases with *SMC*, and the contrast in SAR images between dark flooded and non-flooded terrain is generally larger for moist soils.

The contrast between backscatter from flooded and non-flooded soil can be significantly reduced in the presence of strong wind that roughens the water surface, as schematically depicted in Fig. 2. This is still an open problem for flood mapping from SAR, since radar scattering of wind-roughened surfaces has been deeply investigated only for the open ocean (e.g., [42, 55]). For instance, the relationship between wind (speed and direction) and ocean waves is a function of water depth and *fetch* and is generally defined for well-developed waves in equilibrium with the wind and for deep sea. In flooded terrains, we may encounter many different and unknown situations (different depths, obstacles that screen the wind flow, and so on), where it is difficult to predict the radar signal, even if we know the meteorological conditions accurately. The contrast between flooded and dry surfaces may even completely disappear in extreme cases (high backscatter from high wind-roughened water surface).

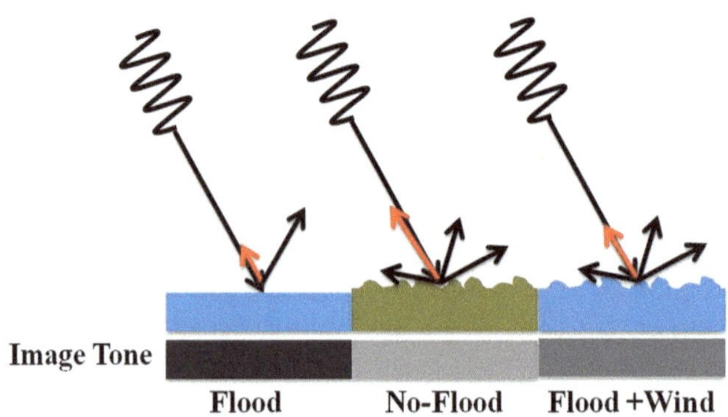

Fig. 2 Scattering mechanisms determining the radar signature of a flood in the presence of variable wind speed that roughens the water surface

In order to quantitatively model the aforesaid mechanisms, the electromagnetic theory of scattering from rough random surfaces must be considered. Closed-form solutions exist for specific ranges of roughness, as summarized, for instance, in Ticconi et al. [59], where many references can be also found. The physics of the problem can be considered quite consolidated, even though the input data needed to use theoretical models for image interpretation are not always easy to gather. For instance, to account for the effect of wind on the contrast between flooded and non-flooded surfaces, one should collect accurate wind speed and direction data over the area of interest. Moreover, targeted areas may become very complex in some cases and difficult to model. For instance, in situations where *SMC* is very high (saturated soil), water ponds can be present within an image resolution cell, producing a mixture of low backscatter from the water ponds and high backscatter from the wet soil, very complex and difficult to predict.

The contrast between floodwater and bare soils is not very much affected by radar frequency and polarization, because soil permittivity and roughness play a similar role for both flooded and non-flooded terrains. Indeed, the soil permittivity law is governed by that of water; consequently, both flooded and non-flooded terrains exhibit a common decrease of permittivity with radar frequency, without a significant effect on the contrast between the two types of targets. Regarding the roughness effect, it must be considered that the roughness dimensions (both vertical and horizontal scales, i.e., roughness standard deviation and correlation length) have to be compared to the radar wavelength. What is important is the ratio between roughness dimension and wavelength. Thus, a surface considered smooth at low frequency (e.g., L band) may behave as a rough surface at higher frequency (e.g., X band). However, this rule applies to both the water and the soil, so that again there is not a significant change in the contrast between the two types of targets. In case of a very smooth water surface, the contrast may be expected to increase at higher frequency, because the soil is equivalently rougher and then brighter in the image. Regarding the radar polarization, it was similarly found that all polarization modes could be used for flood mapping of bare terrains, as the described mechanisms are common to HH and VV polarizations. Although of less importance, the VV polarization is more sensitive to the small waves generated by the wind over a water surface, so that the detrimental effect of the wind for flood detection is mitigated in HH polarization [2]. Indeed, Henry et al. [27] also observed that a better land–water surface discrimination is obtained in HH polarization. Nonetheless, the use of multiple frequencies and polarizations may significantly help in detecting flood. Considering we are always working on complex scenarios, the point here is the possibility of better discriminating among different surface categories, including different non-flooded targets. Working on a multidimensional feature space, each feature bringing different pieces of information, improves the final classification accuracy, including the discrimination of flooded areas from other land cover types. For instance, it has been recently proved that combining HH and VV images can enhance the flood discrimination capability at L-band [32]. This is even more important in case of soil covered by vegetation or built up areas, where the response of different operating frequencies can be very different, as discussed in the following.

2.2 Topography Effects

The models mentioned above generally consider roughness whose dimensions are in the order of or a few times larger than the radar wavelength (wavelengths of 2–10 cm when considering the presently operating spaceborne radars). Indeed, the Earth surface is characterized by its large-scale topography, which is generally known in the form of a digital elevation model (DEM). The effect of the topography can be modeled in a simpler way. The geometric (image deformation) and radiometric (backscattered intensity modulation) effects of the topography are known as foreshortening, layover, and shadowing, in addition to the slant-range to ground-range deformation of a SAR image occurring on a flat surface as well. Here, it is assumed that the reader already knows these basic concepts of radar imaging, which are summarized in Fig. 3. From the point of view of flood mapping, it is worth to point out the occurrence of false alarms due to the presence of shadow. The low level of backscatter (close to the noise floor of the radar system) located on the side of a relief opposite to that of the radar can be confused with the presence of smooth and dark water. The bright pixels on the side of the slope looking toward the radar do not produce errors in this respect, unless the slope is so steep to mix up different parts of the surface in the same pixel (layover), a case that hampers the interpretation of the image. These effects can be predicted if one has an accurate DEM available to orthorectify the image and compute the local incidence angle in each pixel (i.e.,

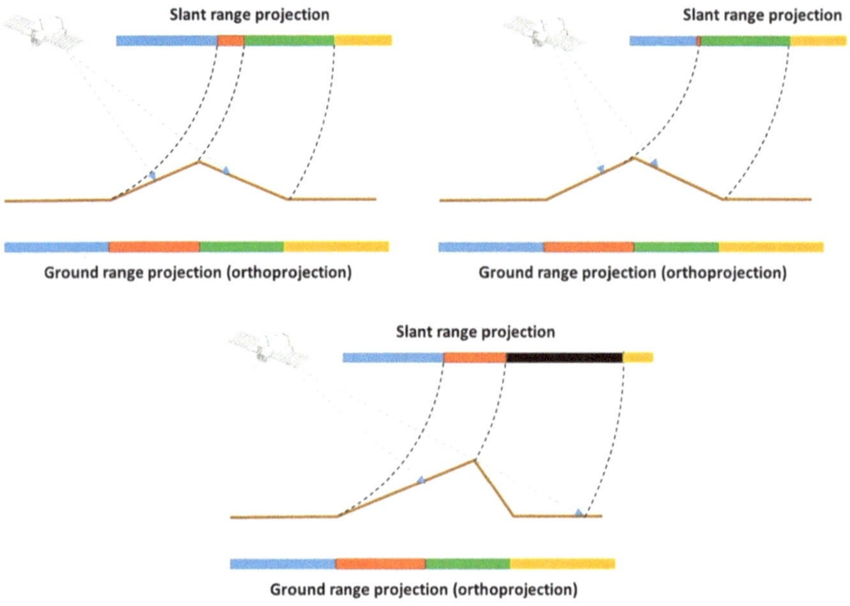

Fig. 3 Foreshorting (*upper left panel*), layover (*upper right panel*), and shadow (*lower panel*) in a side-looking radar image

incidence with respect to the local orthogonal to the surface). Then, a map of areas in shadow, which occurs when the local incidence angle is larger than $\pi/2$ rad, can be computed to avoid false alarms. It is understood that looking at floods in deep valleys surrounded by highly relieved terrains poses some additional challenges for image interpretation. Shadow effects increase with the increase of radar off-nadir angle, so that in case of complex topography, a relatively small incidence angle should be preferred, although if it is too small the layover effect becomes very important.

2.3 Radar Signature from Vegetated Soils

So far, we have compared the radar signature of the bare and rough soil to that of the water. Another important mechanism occurs when flood regards vegetated areas. It is related to the double-bounce mechanism due to the multiple reflection from the horizontal surface and the vertical structures (e.g., the trunk in case of forest or stems for agricultural crops), forming a sort of dihedral type of scattering, as depicted in Fig. 4 in the case of forest. It complements or may even overcome the contributions from single scattering from vegetation, attenuated scattering from the surface, and multiple interactions between the various scattering elements. This mechanism can be highly enhanced by the presence of water under the vegetation, since the coherent specular reflection of the surface increases dramatically in flooded conditions, as discussed before (water has a high permittivity and is usually very flat [29, 41]). This mechanism may occur in the presence of any type of vertical structure, either natural or manmade. It can make the radar return from flooded vegetated or urban areas higher than that under non-flooded conditions, as opposite to what happens over bare soil or any other case in which the water submerges completely the earth surface. The contrast between flooded and non-flooded σ^0 of a forest, that in this case is positive, is however strongly dependent on the characteristics of the vegetation and the observing system.

A major factor, as depicted in Fig. 4, is that the double-bounce enhancement can become the predominant effect when the radar signal is able to penetrate into the vegetation, as in the case of low vegetation biomass, low frequency, small incidence angle, or proper polarization. Another major factor is the predominance of the double-bounce effect in case of HH observations, since the specular reflection from the horizontal surface is much higher for horizontally polarized impinging waves compared to vertical waves (where there is a minimum of reflectivity at the Brewster angle of incidence).

The penetration into the canopy increases at longer wavelengths. This is because, given an extinction process within the media (due to dielectric or ohmic losses, but also to volume scattering in different directions), the attenuation of the wave is a function of a related fraction of the wavelength, and then longer wavelengths go deeper into the media. Consequently, the wave reaches the surface (soil or floodwater) and produces a double-bounce effect when the frequency is low or the vegetation is not very dense and high. It is not easy to predict the occurrence and the intensity of these mechanisms, considering the large number of parameters

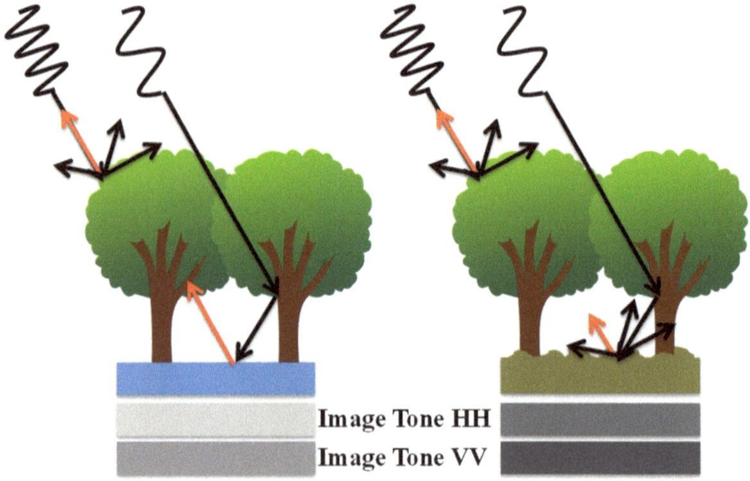

Fig. 4 Double-bounce scattering mechanisms determining the radar signature of a flooded (*left*) and non-flooded (*right*) terrain under vegetation. It shows schematically the larger contribution of double bounce at lower frequency, characterized by higher penetration through the canopy, and the brighter signature of the HH image

that can influence the penetration of the wave and its absorption and extinction through the vegetation layer (e.g., vegetation height, density, aboveground biomass, vegetation water content, shape and orientation of its constitutive elements like branches, leaves, and stems or trunks). In summary, the contrast between flooded and non-flooded vegetation decreases or even disappears if the penetration depth is small and the vegetation single and multiple scattering are the main processes contributing to the radar return. As a rule of thumb, one can consider that the double-bounce scattering enhancement is more important at low frequencies (P and L band). Of course, this is dependent on the density and structure of the vegetation. Then, it might be also observed at higher frequency (say, C and X band, especially at HH polarization) in case of relatively transparent agricultural crops with small elements and low leaf area index (LAI). As for forest, double bounce generally occurs at low frequency only, unless we are looking at deciduous leafless trees. It must be understood that these are not firm quantitative rules to be used when processing images, since the problem is very complex, but they may provide guidelines to address the human image interpretation.

The penetration is also dependent on the polarization, since the vegetation elements elongated parallel to the plane of oscillation of the electric field are those interacting the most with the electromagnetic wave. Generally, horizontal polarization penetrates more than vertical one into vegetation since the orientation of most elements is predominantly vertical. However, as for polarization, the higher specular reflectivity of the water surface in HH is the main reason for preferring this

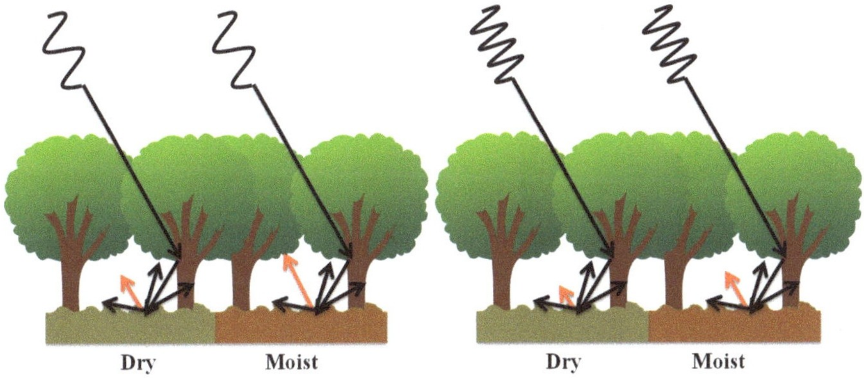

Fig. 5 Scheme of double-bounce mechanisms of radar backscatter from vegetation for different frequencies and soil moisture

polarization to VV when detecting the increase of the double-bounce intensity due to the presence of floodwater under vegetation.

The intensity of the double bounce is also dependent on the incidence angle. Simply speaking, for a dihedral with two orthogonal sides of the same dimension, the highest return is observed at 45°, since this corresponds to the maximum apparent cross section of the target. As for a forest, with almost cylindrical vertical trunks, the height of the trunk with respect to the dimension of the free surface behind becomes important, which contributes to the surface reflection. In addition, it should be considered that the penetration depth is larger for smaller incidence angles, as the path through the layer is shorter before reaching the surface. In summary, it is difficult to predict the angle at which high increase of the double-bounce intensity occurs, but, to detect it, extreme conditions (grazing or steep observations) have to be avoided, and relatively small angles are preferred, at least to guarantee a deeper penetration.

The contrast between flooded and non-flooded vegetation is dependent on the moisture of the soil, although during a flood event one can expect that antecedent rain precipitations produced a wet soil. However, the higher the soil moisture, the higher the double bounce and the soil single scattering, so that the contrast with respect to the flooded vegetation generally diminishes, as depicted in Fig. 5.

Modeling the vegetation σ^0 both in the presence and in the absence of water is a very complex task. It can be solved by computing the solution of the radiative transfer equation (RTE), which assumes that the scattering contributions from different elements of the canopy are statistically independent. Models shall take into account absorption and scattering mechanisms, including multiple scattering (i.e., signal scattered from one element that interacts with another element). The radar return is determined by the statistical properties of the vegetation layer and the ground surface, in terms of electrical (i.e., permittivity) and geometrical (i.e., size, shape, orientation, roughness) parameters of its various elements (i.e., trunks, branches, leaves, ground surface) that determine absorption and scattering volumetric coefficients [7, 60, 61]. In order to introduce a realistic structure of the vegetation, a vertically stratified medium is generally considered, each layer representing different parts of plants and trees (e.g., trunks in the lower layers and foliage in the upper layers). Alternatively, one can rely on simpler models, such as those considering a unique homogeneous layer or neglecting some scattering mechanisms (e.g., multiple interactions between vegetation elements or vegetation–soil interactions). The most popular simplified model is the so-called water cloud model, which assumes the vegetation responds to the radar similarly to a cloud of water droplets [1]. In Fig. 6, the σ^0 computed by a model which assumes the vegetation as a collection of discrete scatterers [7, 18] is reported as a function of *SMC*. Three cases are considered, i.e., a bare soil, a typical forest (biomass of 50 t/ha), and a wheat field (plant height of 70 cm), for different incidence angles. Note that the case of *SMC* equal to 80–100% actually displays the model predictions produced by replacing the rough soil with a fairly flat water surface to simulate the flood. We can see that σ^0 increases with *SMC*, as previously discussed. While the simulations predict a strong decrease of σ^0 for bare flooded soils, a significant increase is predicted for flooded vegetated or forested areas. However, the magnitude of this increase depends on the type of vegetation, incidence angle, and polarization as discussed before (e.g., it is higher at relatively low incidence angle, mainly because of the higher penetration through the foliage).

It is also interesting to understand what happens for a vegetated flooded area as a function of the water depth. At least for agricultural crops, as the water level increases, the phase center of the double-bounce scattering tends to rise and becomes less attenuated by the remaining emergent vegetation. The enhancement of the double-bounce intensity depends on the water depth. For intermediate stage of crop growth, σ^0 increases with water depth, because of the less attenuated double-bounce effect, up to a certain level, when the emerging part of the plants becomes small and is no more capable to generate a relevant double-bounce signal. For a well-developed canopy, the scattering from the vegetation becomes predominant, and the presence of the underlying water is hardly detected by the radar [43].

2.4 Radar Polarimetry and Interferometry

The techniques of radar polarimetry and interferometry deserve their own section. They may represent a step forward for flood mapping by radar. Operationally, the flood mapping application requires a very rapid response of the Earth observation system, and possibly repeated image acquisitions to follow the evolution of the event. These can be feasible only by a satellite constellation (e.g., COSMO-SkyMed or Sentinel 1A/B). The polarimetric radars presently operational (e.g., Radarsat-2, ALOS-2) do not satisfy these requirements, although they can occasionally provide a suitable image after a flood. Nonetheless, radar polarimetry has a great potential to provide a reliable detection of the scattering mechanisms and an

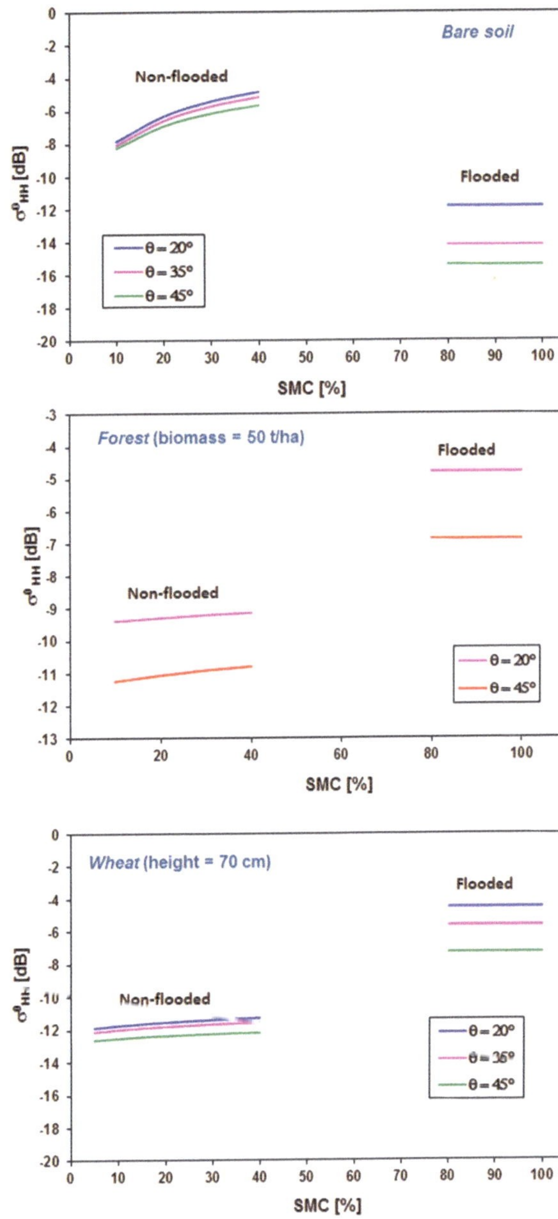

Fig. 6 Backscattering coefficient σ° at horizontal polarization (HH) in dB predicted by the models described in Bracaglia et al. [7] and Ferrazzoli and Guerriero [18] versus the SMC [%] for a bare soil (*upper panel*), a terrain covered by forest (*middle panel*), and a wheat plant (*lower panel*); three different incidence angles (20°, 35°, and 45°) are considered. The simulations performed under flooded conditions are also included (SMC in the range 80–100%) [43]

accurate discrimination of different land covers. Similar considerations apply for SAR interferometry. This technique is based on the acquisition of images with exactly the same geometry (i.e., same incidence angle). A prompt and repeated collection of images during an inundation event may require operating the radar with different angles in different orbits, taking advantage of the steering capability

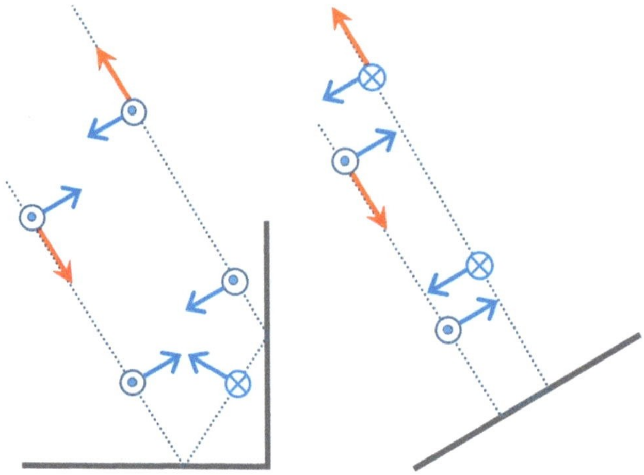

Fig. 7 Simple explanation of the capability to discriminate the presence of a double bounce by looking at the phase between horizontal-horizontal and vertical-vertical radar observations. In case of a dihedral (*left panel*), the received horizontal electric field has the same direction of the transmitted horizontal, whereas the vertical component of the received and transmitted fields are opposite. In case of single bounce from a plate, the horizontal received field is opposite with respect to the incident one, but the vertical component remains as the previous case

of phased array antennas. It is likely that a previous image collected with the same geometry is not available for generating the interferogram. In this respect, the Sentinel-1A/B constellation, thanks to its large swath and short repetition interval, may represent a step forward in this direction.

As far as polarimetry is concerned, it may support flood mapping, especially thanks to the capability to single out the contribution of the double bounce with respect to other volume and surface scattering mechanisms, which helps in detecting its possible enhancement due to the presence of water under vegetation. This capability is due to the ability of a radar polarimeter not only to measure the magnitude of co-polarized (i.e., HH, VV) and cross-polarized (e.g., HV) backscatter, but also to observe the phase difference between them. Simply speaking, it must be observed that the double reflection from a metallic dihedral produces a double reversal of the electric field on the surface (the reflected field is always opposite to the incidence one over a metallic surface as the sum of the total tangential components must be zero). Conversely, a single bounce from an orthogonal metallic plate produces only one inversion. As shown in Fig. 7, the horizontal component of the returned field from a dihedral has the same direction of the incident one, whereas it has the opposite direction in case of the plate, i.e., it has a 180° difference in phase. Then, a radar polarimeter can single out the amount of double bounce produced by the canopy and compare that amount before and after the flood to see if there is an enhancement due to the floodwater [16, 21].

For what concerns interferometry, a major contribution for flood mapping is related to the high sensitivity of the interferometric phase to changes in the arrangement of scatters within an image resolution element (pixel). Indeed the use of the complex coherence is known as a powerful change detection method in many applications [57]. In case of vegetated surfaces, the coherence is always quite low due to temporal decorrelation (e.g., due to wind or plant growing), so that it does not help in detecting the change due to flood water. Conversely, in urban areas the coherence is generally high, but for an interferometric pair with the presence of floodwater in the slave image (i.e., during the flood) and its absence in the master image (i.e., before the flood), a significant decrease can occur, as discussed later on. Finally, we just mention the fact that in case double bounce is the main scattering mechanism in the presence of standing water, the signal can be very coherent in time. Then, it may happen that an interferometric pair with both master and slave images acquired in the presence of floodwater is coherent, and the phase of the interferogram displays interferometric fringes describing the water surface topography or its change in time. Although this is a not very common situation, it can in principle offer an insight into the event from the hydraulic point of view [49].

2.5 Radar Signature from Urban Areas

Radar imaging of urban areas is very complex, especially when the radar resolution is not very high and many scattering components sum up in the same resolution cell. It is therefore currently very challenging to contour the flood in urban areas, which for obvious reasons would be very desirable when managing this kind of disaster. A single building produces a very high backscattered signal due to the double bounce, which originates from the combined reflection of the horizontal ground and the vertical wall facing toward the radar, similarly to what happens in case of trees [20]. The very bright pixels, almost saturating the image in some cases, are located at the side of the building footprint closest to the radar. There are additional scattering components from the building as well. For instance, the layover of the roof return and the ground, which is located at a nearest range with respect to the building footprint, can also produce a high signal partially superimposed to that related to the double bounce (the reader can refer to Fig. 8 to understand that a vertical wall can easily produce a layover effect). Then, there is the return from the other parts of the roof at longer range distance and finally the shadow extending to a range distance larger than that of the building footprint (Fig. 8). Some studies have exploited the dihedral corner reflection to assist flood detection in urban areas, showing the enhancement of the backscatter when the ground is covered by standing water [31, 35]. The complexity of the problem, which includes mutual shadowing between buildings in dense urban settlements, higher-order bounces (triple bounces can be very likely), specular reflection from gamble roofs, and scattering from other element like windows, makes the flood mapping very unreliable. A very powerful technique able to identify changes due

Fig. 8 Scattering mechanisms determining the radar signature of a flooded and non-flooded urban area

to the presence of water is SAR interferometry, and in particular the coherence. As mentioned before, an urban settlement is a very steady target characterized by high interferometric coherence, even for large temporal baselines. The presence of floodwater may imply a significant decrease of the coherence [10, 49]. Then, interferometry offers a unique opportunity to map flood extension in urban areas, although to a spatial resolution typical of the coherence map (size of few pixels required to estimate the ensemble average).

2.6 Rain, Snow, and Other Challenging Conditions

Some surfaces can be sufficiently smooth and have low backscatter (e.g., asphalted roads, flat rock) to be confused with floodwater. Furthermore, there are other situations, not comparably intuitive, which can produce false alarms in a flood detection procedure.

One effect easily observed during a flood event, when heavy precipitations may occur, is the atmospheric attenuation of the radar signal. The amount of absorption (water is a highly dissipative medium) and scatter (a function of drop size) of the radar signal due to water drops is higher at higher frequencies. Therefore, a high attenuation from heavy rain can be observed in the images collected at higher frequency, such as X-band. This produces very low backscatter that can be misinterpreted as floodwater [47]. This problem is not severe at lower frequencies (C and especially L band) but can be important at X band if one discriminates the flood just using the image intensity.

Another effect that can cause false alarms in flood mapping if not correctly detected is the attenuation due to wet snow. The snowpack is quite transparent in the microwave spectral range when it is dry, as ice is not much dissipative (ice particles scattering can be observed at higher frequencies). When the snow is wet (i.e., there is a certain amount of liquid water), similarly to the case of rain, it becomes a very absorbing material that produces very low backscatter and can therefore be misinterpreted as floodwater [47]. In this case, one can rely on ancillary information to avoid committing false alarm errors. For instance, it is generally expected that in flat areas (e.g., valleys in case of mountainous areas), the information on topographic slope and local incidence angle can be very useful to mask areas where very low backscatter is likely not due to the presence of floodwater.

3 Flood Detection Methods

In this section, we summarize algorithmic approaches that have been used to map flood in three main land cover classes, such as bare soil, vegetated, and urban areas. The more challenging sites are still urban and vegetated areas, since many variables concur to characterize the backscattering signal, albeit some issues are still open for what concerns bare soil areas. In particular, here we tackle all those stimulating situations where accurate flood mapping can require proper strategies based on the acquisition of more informative data sets, either from satellite or from ancillary data sources supported by models. The possibility to exploit mission data provided by radars working at different frequencies and/or polarizations represents a unique opportunity to improve the accuracy of flood mapping. The increasing accessibility of images systematically acquired by new satellite constellations (i.e., Sentinel-1) represents a step forward in this application. Indeed, the availability of images with short repeat time is very powerful, especially when supported by a good knowledge of the land cover and related scattering mechanisms. Exploitation of InSAR coherence is another useful tool, given that an interferometric pair of images is collected during the flood event.

3.1 Bare Soil

The majority of SAR-based flood mapping algorithms deal with the detection of water in bare soils or scarcely vegetated areas, using approaches that are quite well established; they provide maps that are reliable and frequently used for emergency services [54]. The algorithms to detect flood in bare soils can be split in two main families, the ones that make use of just one image, i.e., the flooded one, and those that take advantage of one image more, i.e., the reference image, which usually observes the same region but in dry conditions. The first group of algorithms searches for pixels with low backscatter values, associable with a high probability of the presence of standing water, while the others detect the decrease of backscattering with respect to a dry condition image, making use of change detection approaches. Both approaches have advantages and drawbacks. Indeed, on the one hand the use of just one image reduces the processing time and avoids relying on another satellite acquisition, which is quite advantageous especially with satellites such as COSMO-SkyMed for which an interferometric reference acquisition is generally not available. On the other hand, false alarms may occur because there are land cover classes that have backscatter values comparable with water or even lower, such as smooth surfaces (flat roofs or asphalted areas), very dry regions (deserts), and shadowed regions. The latter false positives can be removed in case an interferometric reference acquisition is available, using change detection approaches, since the geometry is exactly the same in both images. However, the shadow area remains a region where we do not have any information; thus hydraulic models or other ancillary data can be used to fill the gap [41, 48]. Interferometric acquisitions are images acquired from the same orbit, with identical looking angle and just separated by a perpendicular baseline, which can reach hundreds of meters. Thus, in case of change detection approaches, interferometric acquisitions are preferable because the detected changes can be directly ascribed to the occurred changes on the surface, so dispelling the possibility of being due to changes on incidence angle and avoiding using surface scattering models to correctly interpret them [44].

Using only the flood image and reducing the flood mapping issue to a classification problem between two classes, i.e., water and non-water, the backscattering thresholding is undoubtedly the fastest and most frequently applied technique. Generally, the threshold is identified as the maximum backscatter value associated with open water, and consequently, all pixels having an intensity value that is lower than the fixed threshold are classified as inundated [33]. Different approaches can be used to derive these thresholds, subjective visual inspection, or automatic procedures depending on global (one threshold for the entire image) or on local information (one threshold for each subregion of the image). Parametric algorithms are generally used to automatically extract the best threshold value, although their efficiency is known to be strongly hampered when the respective fractions of the image occupied by the different classes are strongly unbalanced or when the distribution functions significantly overlap due to the presence of the speckle noise. To overcome these difficulties, tiling approaches with fixed [5, 34] or variable [15] tiles size are needed. With this kind of classification algorithms, the uncertainty is higher in the portion of the histogram where both distribution classes overlap [23]; thus, it is exactly in this region that algorithms have to find an efficient solution to avoid high false alarms rates. Indeed, a thresholding algorithm inevitably produces a binary map with a number of misclassified pixels that depends on the extent of the overlapping distributions. Several approaches have been proposed to deal with the issue of the overlap region, such as (1) combining the Otsu thresholding method on tails of fixed size with region growing [47]; (2) using thresholding, tailing, and fuzzy logic in the post-processing phase [34]; (3) integrating thresholding, change detection, and region growing [36]; and (4) hierarchical tiling and region growing [15]. Statistically based approaches have been also directly applied to solve the two-class problem in the image difference using an expectation–maximization algorithm with generalized Gaussian distributions [3] or using a support vector machine and a similarity measure [6].

Although the detection of flood in bare soil is quite straightforward, some unsolved issues and some false alarms may still be present. In this respect, soil moisture plays an important role, increasing the backscattering value with respect to dry conditions, whereas flooded vegetation gives rise to the double-bounce effect (water surface-tree trunks). Ancillary data and land cover maps can be fruitfully used in order to disentangle these issues, but also the InSAR coherence can add further information content to the analysis based only on the incoherent part of the backscattering. In this respect, in the case of the Tsunami in Japan on 2011, the InSAR coherence was used to interpret the increase of backscatter occurred between two COSMO-SkyMed acquisitions with a time span of 1 day in areas where the flood extent was receding toward the shore [10]. In these areas, the receding process led to an increase of the backscatter from one day to the next, which is similar to the effect of an increasing soil moisture content. The large value of InSAR coherence between the two acquisitions indicated that there was not any change in the spatial arrangement of the scatterers on the terrain, so that the variation of backscatter has been ascribed to changes in soil moisture. In this particular case of an inland flood caused by a tsunami, the InSAR coherence has been used also to detect another secondary effect caused by the big earthquake, such as the liquefaction occurred in urban areas [11]. Although this phenomenon has a relatively small magnitude, it occurs frequently in case of a seismic event and is always easily detectable by the InSAR coherence [12].

Of course, having at disposal a long time series of SAR data, the seasonality of the backscattering can be modeled, thus characterizing the seasonal behavior of the soil moisture and land cover. To this purpose in Schlaffer et al. [51], harmonic analysis of a multitemporal time series of more than 500 ASAR scenes with a spatial resolution of 150 m was used to characterize the seasonality in backscatter under non-flooded conditions. In this study, pixels that were inundated during a flood event showed strong deviations from the seasonal trend inferred from the harmonic model. The residuals, i.e., flooded pixels, were classified by means of an automatic threshold optimization algorithm, after masking out areas which were unlikely to be flooded using a topography-derived index. Furthermore, in Schlaffer et al. [52], multitemporal C-band ASAR data have been used to retrieve harmonic model parameters of permanent water bodies, seasonal open water, persistently flooded vegetation, and seasonally flooded vegetation. They were used to classify wetland areas at regional scale, such as the case of the Kafue River basin in Zambia, where a time series of around 100 SAR images spanning a time period of almost 2 years has been used. Nowadays similar approaches, although computationally expensive, could be made possible by the availability of frequent and systematic acquisitions provided by the Sentinel-1 ESA satellite constellation.

3.2 *Vegetated Areas*

Double-bounce backscattering represents the key process to detect flooded vegetation. Indeed, when the ground is covered by a smooth and very reflective water surface, the intensity of the double-bounce effect involving the surface and vertical structures such as trunks is increased, if the penetration through the canopy is sufficient. This increase can make the backscatter from flooded vegetation higher than that from non-flooded vegetation, so that the former may appear brighter in a SAR image. However, flooded vegetation does not always have a clear radar signature that can be easily detected, and frequently a support from vegetation scattering models is necessary, especially in the change detection approaches, if interferometric acquisitions are not available [44]. It is worth pointing out the importance to account for speckle noise, which generally prevents the comparison with modeled backscattering on a pixel basis. The application of standard multi-looking techniques to reduce the speckle effect prevents the full exploitation of SAR image spatial resolution, penalizing very high-resolution sensors, such as COSMO-SkyMed, which have the potential for a detailed mapping of the landscape, i.e., an accurate flood boundary delineation. A possible efficient way to deal with speckle is the segmentation approach, as done by Pulvirenti et al. [44], where the image segmentation has been accomplished by using the morphological profile by open and close filters [39]. The morphological operators were applied to a multitemporal series of five COSMO-SkyMed images in order to extract information on the spatial extension of homogeneous objects present in the scene. The morphological profiles carry information at different spatial scales, thus singling out the radar response directly related to the targets that have to be classified and coping with the speckle noise in radar data. It is important to highlight that the signature of each segment, which represents the main information derived from the segmentation procedure, was almost speckle free, since the averaging process within each segment drastically reduces the speckle noise. Morphological operators are widely exploited in very high spatial resolution images, both SAR and optical, to extract contextual information for supporting urban classification [4, 9, 44], and the basic operators are erosion and dilation [56]. Generally, they are applied to images using structuring elements (SE) of known shape, corresponding to specific geometrical structures, to depict how structures in the image match those specific elements. Two other morphological operators obtained from the previous ones are the opening and closing operators. The function of opening is to dilate an eroded image in order to recover as much as possible of the original image, while the function of closing is to erode a dilated image in order to recover the initial shape of image structures that have been dilated. The characteristic of the filtering process provided by the opening and closing operators is that not all structures within the original image are recovered when these operators are subsequently applied. The size of the SE, with respect to the size of the structures actually shown in the scene, conditions the output of the filtering operation. In fact, some structures in the images may have a high response for a given selected size and a lower response for other sizes. In particular, this technique is used to isolate bright (opening) and dark (closing) structures in the image. Moreover, dealing with high-resolution data, we generally have a reduced number of scatterers per resolution cell, meaning that the speckle cannot be considered fully developed [50]. Thus, no assumption, or unreliable ones, can be done on the distribution of speckle, a reason to prefer nonparametric segmentation approaches.

When fully polarimetric data are available, double-bounce scattering in vegetated area is easily detectable by the phase information using polarimetric decompositions [16, 21], although using only co- and cross-polarized channels is still possible to identify water beneath vegetation. Indeed, the cross-polarized backscatter is more sensitive to volume scattering, while the co-polarized one is more sensitive to the double bounce. As a matter of fact, in vegetated areas we usually detect a sensible backscatter increase in the co-polarized channels, and much less in the cross-polarized one [28]. In Chini et al. [14], the two polarizations of Sentinel-1, i.e., VV and VH, have been exploited to detect the floodwater under vegetated canopies for a test case in Greece. The algorithm made use of change detection methods and was composed of three main steps:

1. Detection of floods on bare soil by exploiting the expected decrease of VV backscattering;
2. Labeling as potentially flooded of the regions that were not been identified as flooded in the previous step and that showed an increase of the backscattering in the VV channel;
3. Segmentation of the map provided by the step 2 and computation of the mean value of the VH backscattering for each segment;
4. Labeling as flooded of those segments that did not show any increase of backscattering in VH.

The identification of the increase and decrease of backscattering has been performed using an automatic parametric thresholding algorithm developed by Chini et al. [15], which makes use of a hierarchical split-based approach (HSBA). The hierarchical tiling of the scene is carried out in order to automatically select tiles where the statistical distribution function attributed to water, and changes of backscatter values due to the appearance of surface water, can be parameterized. Based on the fitted distribution functions, in the second step an optimization procedure is run in order to extract the optimal threshold to select the seed pixels of the subsequent region growing and change detection steps.

Moreover, vegetated areas are usually characterized by volume scattering, which determines low coherence (even for short temporal baseline) due to changes in plant growing stage and movement of stalks and/or leaves caused by wind. Consequently, coherence can help in distinguishing increases of backscatter due to soil moisture variations (bare soils) from those due to the presence of floodwater (double-bounce effect in vegetated fields), where InSAR coherence is considerably lower. In this case, the coherence has basically the role of replacing optical data to detect vegetation, provided that the spatial and temporal baselines are small enough to assume that spatial and temporal decorrelations do not alter coherence values in the absence of vegetation [49].

Change detection approaches have been also applied making use of fuzzy theory, which is discussed more in detail in the following, to integrate SAR measurements, before and after a flood event, with additional data, such as land cover maps and digital elevation models (DEMs). The information on land cover permits to account for both specular reflection, typical of open water, and double-bounce backscattering, typical of forested and urban areas, while a DEM was useful for including simple hydraulic considerations on the dependence of inundation probability on surface characteristics. Contextual information was taken into account as well, in the post-processing phase to remove false alarms or to include missed pixels [41]. A Bayesian network was also proposed by D'Addabbo et al. [17] to integrate remotely sensed data, such as multitemporal SAR intensity images and InSAR coherence data, with geomorphic and other ground information to detect flood in vegetated areas.

3.3 *Urban Areas*

Franceschetti et al. [20] explained that the double-bounce effect in urban areas is typically represented in SAR images by a very bright line that appears on the side of the wall that is illuminated by the radar. It is worth mentioning that double-bounce paths have the same length as the range of the intersection between the ground and the front wall of the building. In this case, even if the building is surrounded by smooth asphalt surfaces, so that the change in soil roughness may be negligible, the increase of reflectivity due to the increase of the dielectric constant implies, in principle, a considerable increase of the double-bounce effect. Stemming from this principles, Mason et al. [35] investigated the possibility to detect urban flood looking at the layover regions, estimating double scattering strength using one TerraSAR-X image (3 m resolution) in conjunction with a high-resolution LiDAR height map of the urban area. They also used a SAR simulator to generate maps of layover/shadow from LiDAR data, to estimate the position of double scattering ridges in the SAR image. Observations of double-bounce backscattering strength were compared to the predictions from an electromagnetic scattering model to identify flooded areas.

The possibility to measure this increase of backscattering in the double-bounce area highly depends on the geometric accommodation of the building façade with respect to the line of sight of the SAR sensor, resulting in higher or lower sensitivity to backscattering change magnitude. Indeed, if we consider the angle between the flight direction and the street alignment, the increase is very high at 0° (i.e., walls parallel to the azimuth direction) but is reduced for larger angles. Stemming from the model results of Ferro et al. [19], which assumes for simplicity an isolated building, Pulvirenti et al. [49] found that for angles greater than 5–10°, the increase of the backscattering is in the order of 3.5 dB. As a consequence, even in an ideal situation of an isolated building surrounded by a homogenous ground surface, the increase of the double bounce due to the presence of floodwater is not very high if buildings are not parallel to the SAR flight direction. This increase of backscattering can be hardly detected taking into account only the intensity, especially in a complex urban environment where the assumption of isolated buildings is often unrealistic. Hence, the idea to couple intensity data with other features extracted from SAR data, such as the coherence, has been developed using COSMO-SkyMed images for the Japan tsunami event on 2011 (stripmap mode, 3 m resolution, [10]). For that event, images with a temporal baseline of 1 day were available, and some areas showed a stable value of intensity within the double-bounce ridges, combined to a very low value of coherence. Low coherence in urban areas affected by a seismic event, as the Japanese earthquake, is generally due to damage in manmade structures (e.g., [8, 9]). In that particular case, the interferometric pair was a post-seismic one, characterized by short temporal (24 h) and spatial (10 m) baselines, so that the low coherence values could not be ascribed to collapse of buildings, or to spatial decorrelation [65], but to the presence of floodwater. The presence of water between buildings was confirmed by optical images available over the area.

The interferometric coherence measures the degree of correlation between two complex (phase and amplitude) SAR images. It is particularly related to the change in the spatial arrangement of the scatterers within a pixel [12] and thus to geometric changes in the scene. An interferometric pair can be built using two images taken before the flood (hereafter denoted as pre-event pair), after the flood (post-event pair), or one image before and another after the flood (co-event pair). It is expected that flooded areas exhibit low coherence that help distinguishing them from non-flooded regions, especially over targets where this feature is expected to be high. Urban areas are generally considered as coherent targets [58]; thus the temporal baseline should not affect the interpretation in a significant way in a stable urban environment; however, it must be verified that a loss of coherence is not due to an increase of the perpendicular baseline, which is another parameter that can affect its magnitude. Indeed, the increase of the spatial baseline and, namely, its perpendicular component implies a decrease of the coherence, thereby reducing the capability of this feature to help discriminating changed and unchanged urban targets [65]. The temporal baseline is in some way related to the changes in the scene. For this reason, we should privilege pairs of images with short temporal baselines, especially in vegetated areas. In an urban area, there is no such a strict constraint, because the presence of vegetation is rather limited.

The role of the interferometric coherence in complementing intensity SAR data for mapping floods in urban areas has been further investigated in Pulvirenti et al. [49], where a proof of concept on how to integrate the two features was developed. They firstly discussed the advantages of using intensity and coherence from a theoretical point of view and then verified the hypothesis for the flood event that hit the Emilia–Romagna region (Northern Italy) in January 2014 and was imaged by COSMO-SkyMed satellites. The short revisit time of the COSMO-SkyMed images, as well as a dedicated acquisition plan tailored to the requirements of the Italian Civil Protection Department, allowed a data set of radar interferometric observations of the event to be collected. Results showed that the analysis of the multitemporal trend of the coherence was useful for the interpretation of SAR data since it enabled a considerable reduction of classification errors that could be committed considering intensity data only. Not only the InSAR coherence but also the classical Differential SAR Interferometry (DInSAR) technique was used to measure the water-level variation in vegetated areas where the water was receding.

It is worth to highlight that changing the spatial resolution changes the building archetype in a SAR image. Indeed, with metric resolution images (e.g., 1 m), the individual scattering mechanisms of a building can be detected and singled out, such as layover, double-bounce (for which the presence of the water is expected to have a role), shadow, and roof direct return. With decametric resolution (e.g., 20 m), the contributions are mixed in the same resolution cell [13], thus the information about changes can be extracted more at regional scale rather than for each building. In this respect, in Chini et al. [14], the capabilities of the InSAR coherence to detect water in urban areas have been also showed with lower-resolution images (i.e., 20 m), such as the ones provided by the Sentinel-1 satellite mission. The test case was a flood event that occurred in Sicily (Italy) on November 2015, and the dataset was composed of three Sentinel-1 images. Two of those were acquired before the event, and one was obtained during the event. To highlight the presence of water in urban areas, two coherence maps were computed, one pre-event and one co-event. Their mean values at the scale of villages/cities were calculated, and villages affected by the flash flood event showed a decrease of coherence in the co-event one, while for those villages far from the event, the coherence remained stable.

Gaps still remain in urban areas, such as shadowed areas or complex urban structures where SAR backscattering does not provide any clue about the possible presence of water, and they could be solved by the assimilation of the flood maps into hydraulic models [22, 30, 63]. For the assimilation procedures, the flood maps should be provided in terms of probability of a pixel being flooded [26, 53], in order to characterize all uncertainty in flood mapping delineation [25]. Moreover, the direct assimilation of probabilistic flood maps into hydrodynamic models is beneficial because it eliminates the intermediate step of extracting water levels from a binary flood extent map. Additionally, models enable a more accurate and reliable short- to medium-range prediction of flooded areas and provide indications of other related hydraulic variables, such as water elevation, river discharge, flood duration, and flow velocity. If used as inputs to different types of impact models, these variables can better assess the economic, social, and environmental consequences of floods. Indeed stakeholders in flood management are particularly interested in urban areas and all the information characterizing the areas in terms of hazard [24].

4 Flood Detection Examples

In the previous sections it was pointed out that the simple assumption that floodwater is a smooth surface that behaves as a specular reflector giving low backscatter does not account for complications in SAR imagery due to the presence of vegetation or urban settlements. It was also underlined that, even in an ideal case in which flood involves only bare or sparsely vegetated soils, speckle noise generally prevents image classification using only the backscattering value on a pixel basis. Hereafter, we present some examples of operational or semioperational algorithms, in which flood mapping in vegetated and urban areas is accomplished. The bare soils case is tackled as well, given that it is always accounted for in any flood mapping algorithm. The presented examples, developed by the authors, make use of the fuzzy logic as a decision tool, pixels or objects as the unit to extract meaningful information, and magnitude of SAR changes, both intensity and InSAR coherence, as sources of information to classify an area, eventually compared with models.

4.1 Vegetated and Urban Areas: Fuzzy Logic and Pixel-Based SAR Intensity

An image processing technique that can be considered very useful for flood mapping is the fuzzy logic [64]. In fact, a fuzzy approach intrinsically accounts for the various uncertainties inherent to the detection of floodwater, especially in vegetated terrains and urban settlements. Pierdicca et al. [41] firstly applied it for flood mapping purposes. The fuzzy sets represent an extension of the classical notion of set. While in classical set theory an element either belongs or does not belong to the set, elements of a fuzzy set have degrees of membership. These degrees are described by a membership function whose values are real numbers in the interval [0–1]. The fuzzy theory is suitable for representing the sets for which the definition of a membership criterion is a difficult task. This is the case of the set of flooded pixels in SAR images. The fuzzy logic enables to combine several pieces of information in a classification algorithm in order to improve the reliability of maps of inundated areas. In particular, it is useful to integrate in the classification procedure prior information (e.g., a DEM of the area hit by the flood, land cover maps), simple hydraulic considerations (e.g., the dependence of inundation probability on surface slopes or the nearness to permanent water bodies), and contextual information.

A first example of application of a flood mapping algorithm based on fuzzy logic concerns the event occurred in the Alessandria district (Northern Italy) on November 5–6, 1994 [41]. A pair of ERS Precision Images (PRI), collected before (October 3, 1994) and after (November 9, 1994) the inundation, were used, together with a DEM of the area and a land cover map (CORINE land cover). The flood image is shown in Fig. 9, where not only dark areas indicating the presence of floodwater in bare soils but also very bright spots indicating the occurrence of intense double-bounce effects are clearly visible.

The classification scheme applied to the dataset collected to map the Alessandria flood is illustrated in Fig. 10. It can be seen that five fuzzy sets are generated. For the dark pixels (areas of low backscatter), the fuzzy set is defined through the standard Z membership function (Pal and Rosenfeld [37]), for which the lower the intensity of the pixel, the higher the degree of membership. More interesting for the topic of this chapter is the fuzzy set of high difference between the backscatter of flood and pre-flood images. To define this set, which encompasses the areas where a double-bounce effect takes place (e.g., vegetated flooded areas) or increases its intensity due to the presence of floodwater (e.g., urban areas), the standard S function (i.e., 1–Z) is used; standard S and Z functions are shown in Fig. 11. The values of the parameters of these functions (x_1 and x_2; see Fig. 11) can be chosen based on a

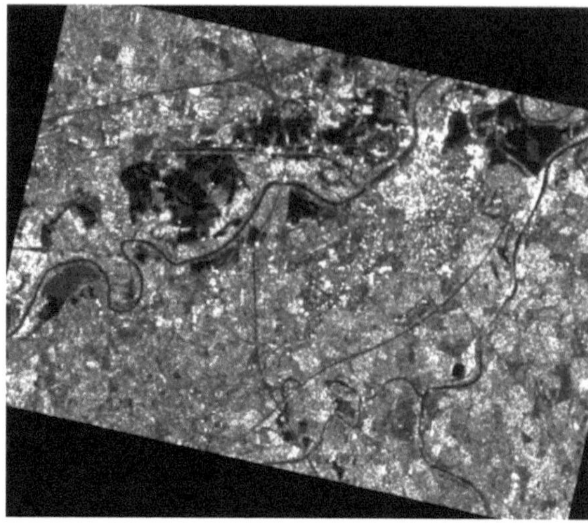

Fig. 9 ERS-1 image of the Alessandria (Italy) flood acquired on November 9, 1994 (From Pierdicca et al. [41], © 2008 by the authors)

supervised analysis of the available data (especially if some ground truth data are available) or based on other kinds of data, such as the predictions of models able to simulate the radar return from floodwater (see the following).

Looking at Fig. 10, the use of the fuzzy logic to complement the information brought by SAR (both pre-flood and flood images) with ancillary information, such as that provided by a DEM, can be noted. In this example the fuzzy rules based on the DEM consider that there is a high probability of flood occurrence in low altitude areas, in flat areas, as well as in concave areas [41]. Hence, in the scheme of Fig. 10, the DEM is processed to compute slope and concavity, the latter being derived from the DEM Laplacian. The different fuzzy sets are then combined through the fuzzy union (i.e., computing the maximum membership degree) and the weighted average of the degrees. Then, contextual information is taken into account through a correction of the membership degree of each pixel based on the statistics of the degree of membership of the neighbors. Finally, the so-called defuzzification process transforms a fuzzy number into a crisp number. It is performed by simply classifying as flooded the pixels having a final membership degree greater than 0.5. The result of the application of the fuzzy algorithm described above for the Alessandria case study is shown in Fig. 12.

The accuracy of the map shown in Fig. 12 is assessed using data derived from a ground survey performed when the inundation reached its maximum extension [41]. Since the SAR image of the flood was acquired 3 days after the flood, only the commission errors (false alarms) can be evaluated. 87% of the pixels are correctly classified as flooded, and a 5% of false alarm is obtained (the residual 8% corresponds to rivers). It is interesting to note that through the fuzzy algorithm accounting also for the increase of the backscatter due to the double-bounce effect, even pixels located in the urban area of the city of Alessandria were correctly classified as flooded.

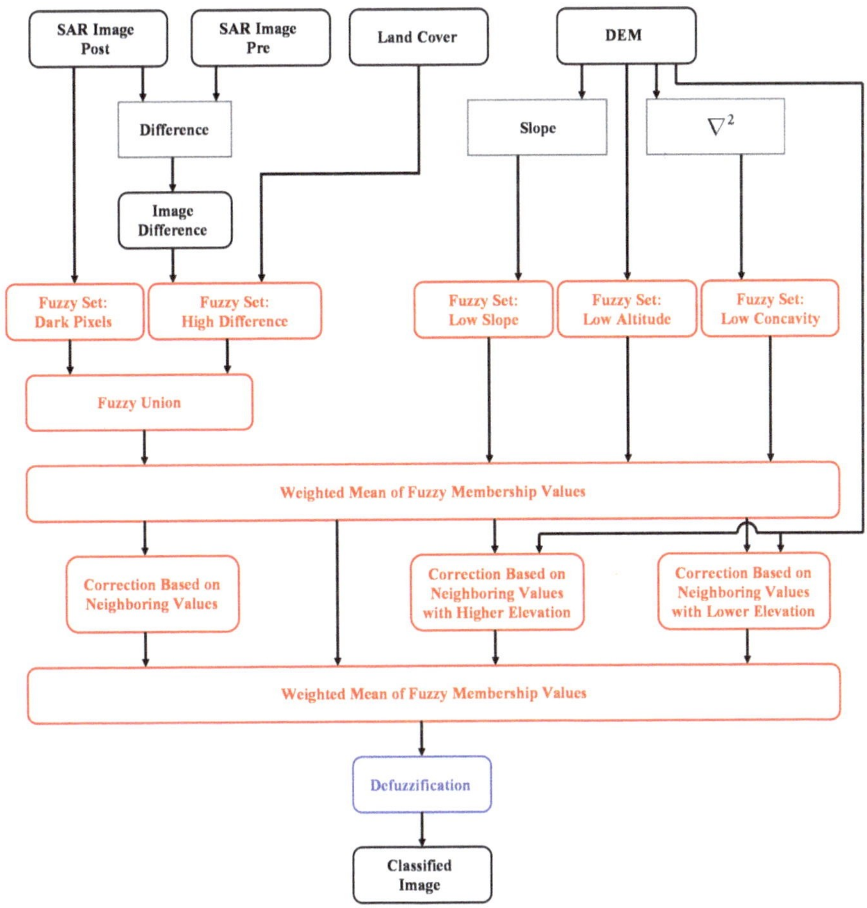

Fig. 10 Block diagram of a fuzzy logic-based flood mapping algorithm (From Pierdicca et al. [41], © 2008 by the authors)

4.2 *Vegetated and Urban Areas: Fuzzy Logic, Pixel-Based SAR Intensity and Models*

As mentioned above, since in most of the cases ground truth data are not available to precisely set the values of the parameters x_1 and x_2 of the fuzzy membership functions (Fig. 11), one can rely on the simulations of models able to predict the radar return from flooded surfaces. An example is represented by the models developed at the University of Rome Tor Vergata [7, 18] to simulate the radar return from agricultural and forested areas and proposed by Franceschetti et al. [20] to predict the backscatter from urban areas. Pulvirenti et al. [45] and Pulvirenti et al. [46] used these models to map floods from very high-resolution X-band COSMO-

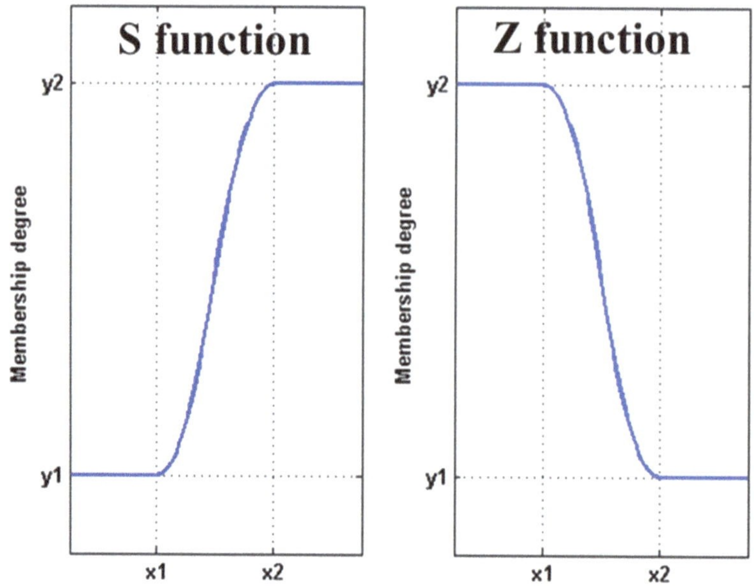

Fig. 11 Standard S and Z fuzzy membership functions

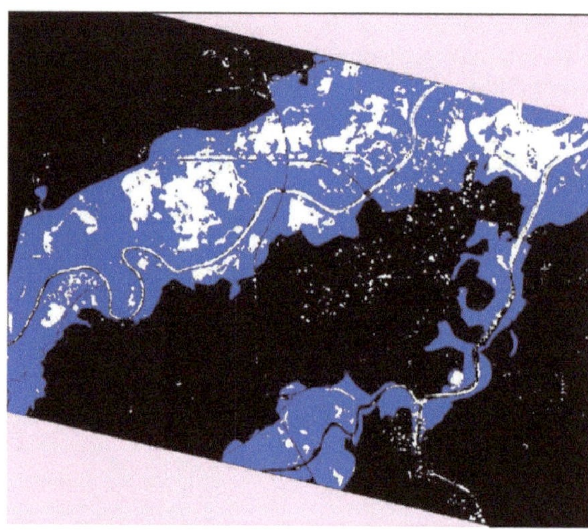

Fig. 12 Map of the flooded area (*white* regions) derived from the application of the fuzzy algorithm shown in Fig. 10 to the ERS-1 images of Alessandria (1994, Italy) shown in Fig. 9. Note that that the *large blue* areas (flooded according to a ground survey) not covered by *white* ones do not imply that the algorithm underestimates the inundation. They are due to the temporal difference between ground survey and the SAR image of the flood (From Pierdicca et al. [41], © 2008 by the authors)

SkyMed data. An example of model-derived parameters of the fuzzy functions is reported in Table 1 for X band, incidence angle θ equal to 35°, and horizontal polarization. The values in Table 1 are suitable for the standard Z function (low backscattering areas) and the standard S function (agricultural terrains, urban areas, forests), which takes into account changes of backscattering.

Table 1 Fuzzy thresholds derived from the electromagnetic models used by the authors for X band, incidence angle equal to 35°, horizontal polarization

Class	x_1 [dB]	x_2 [dB]
Low backscattering areas	−19	−10
Agricultural terrains	2	4
Urban areas	4	6
Forests	3	5

From Pulvirenti et al. [44], © Author(s) 2011

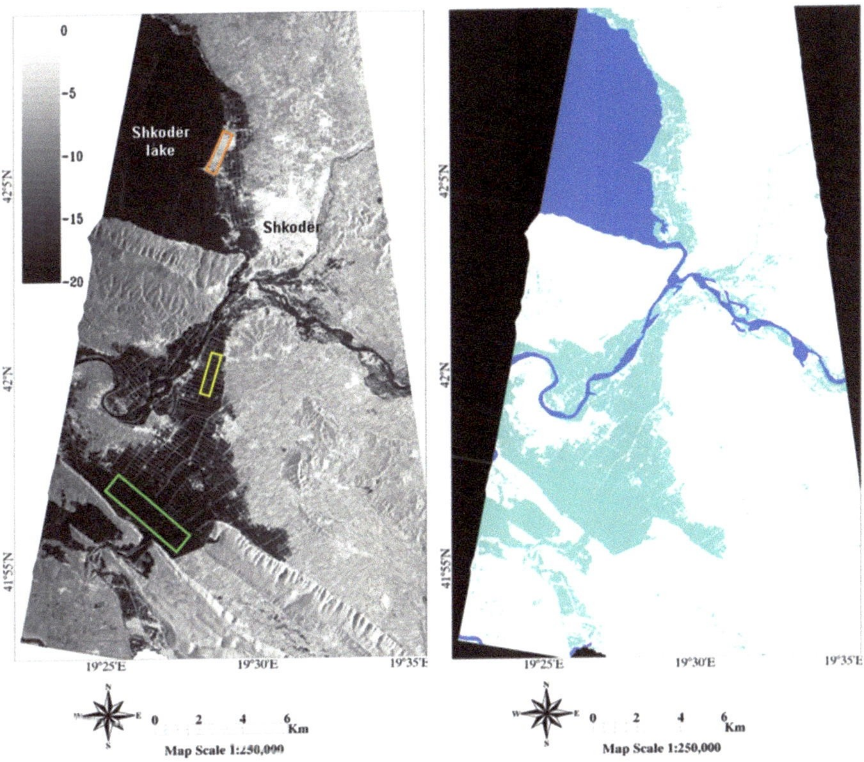

Fig. 13 COSMO-SkyMed Image of the Shkodër (Albania) flood acquired on January 10, 2010 (*left panel*) and result of the application of the fuzzy algorithm (*right panel*). For the *right panel*, *white*, non-flooded; *cyan*, flooded; *blue*, water bodies according to the land cover map; *black*, no data (From Pulvirenti et al. [44], © Author(s) 2011)

Examples of flood maps obtained through a fuzzy logic-based algorithm using the values reported in Table 1 are shown in Figs. 13 and 14. They concern the flood that occurred in Albania, close to the city of Shkodër, in January 2010 [45]. It is interesting to note that despite the use of X band, which is not favorable to detect floodwater beneath vegetation, a forested area (identified by the orange box in the left panel of Fig. 13) is classified as flooded (right panel of Fig. 13). Even though for the 2010 Albania flood validation data are not available, this result can be considered as reliable. In fact, the forested area is adjacent to zones that appear very dark in the

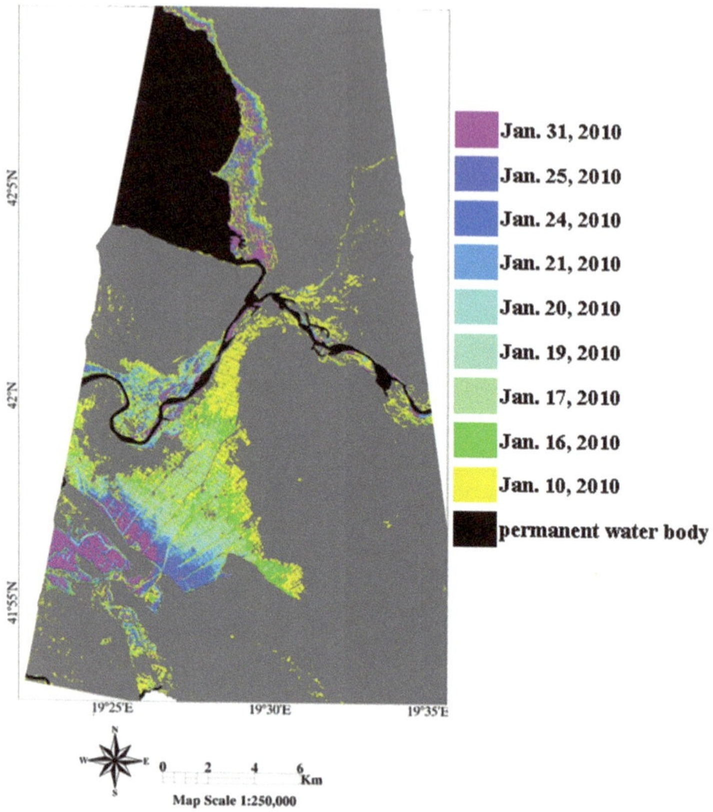

Fig. 14 Multitemporal map showing the time evolution of the January 2010 Shkodër (Albania) flood. It has been produced by applying a fuzzy logic-based algorithm, trained by backscatter simulations, to nine COSMO-SkyMed images whose acquisition dates are reported in the legend. *White* no data, *gray* non-flooded (From Pulvirenti et al. [44], © Author(s) 2011)

SAR image (see Fig. 13, left panel), which can be therefore assumed as flooded. Since the DEM of the area does not show significant changes in the height of the forested area with respect to the surrounding inundated zones, there is no physical reason to suppose that only this area is not covered by water.

Figure 14 aims at demonstrating the utility of recently launched constellations of satellites hosting SAR sensors (COSMO-SkyMed, Sentinel-1) to monitor the temporal evolution of a flood. In particular, it shows a multitemporal map representing time evolution of the Shkodër flood; the gradual receding of water can be clearly observed. The short revisit time of COSMO-SkyMed data enables to produce flood maps with an unprecedented temporal resolution.

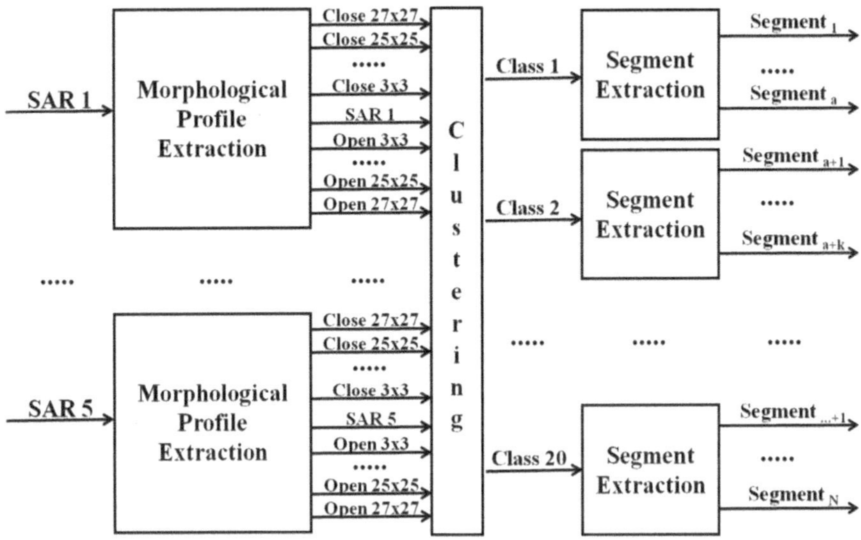

Fig. 15 Block diagram of a segmentation procedure based on mathematical morphology (From Pulvirenti et al. [45], © 2010 Elsevier Inc. All rights reserved)

4.3 *Vegetated and Urban Areas: Fuzzy Logic, Object-Based SAR Intensity and Models*

In the previous examples, flood mapping algorithms based on fuzzy logic have been applied in the framework of a pixel-based classification. To better cope with the speckle noise, it is possible to apply the fuzzy logic, as well as any other flood mapping algorithm (even simple image thresholding), in the framework of an object-based approach, i.e., after having performed a segmentation of the SAR images [44, 46]. An example of segmentation procedure that can be applied to multitemporal SAR images is shown in Fig. 15.

The procedure schematized in Fig. 15 is based on mathematical morphology and uses the opening and closing operators. In this example, a multitemporal dataset of five SAR images is used, but the scheme can be easily adapted to any number of SAR images composing a multitemporal time series. The opening and closing operators are applied to the SAR intensity images, using square structuring elements with different sizes ($3 \times 3, 5 \times 5, 7 \times 7, \ldots, 27 \times 27$ for this example). In this way, a multitemporal morphological profile is created, which, for the example illustrated in Fig. 15, is composed by 27 elements for each SAR image (13 element for the closing operator, 13 elements for the opening operator plus the original backscattering image), for a total of 135 inputs (27 elements $\times$ 5 days). After the input space is created, a K-means clustering algorithm is applied to the set of 135 features, in order to account for the contextual information, about the extension of objects brought by the morphological profile and for the evolution in time of the intensity through the five available SAR acquisitions. In the example in Fig. 15, it is assumed that 20 represents a reliable estimate of the number of classes distinguishable in the observed scene. The spatial contiguity is considered in the last step of the segmentation procedure, which is the class disaggregation into segments. The pixels included in each segment belong to the same class and are contiguous in space.

An example of application of the morphological segmentation to a time series of COSMO-SkyMed images is shown in Fig. 16. The time series refers to the flood that occurred in Tuscany (central Italy) in December 2009. In particular the SAR images were acquired on December 20, 2009 (pre-flood), December 30 and 31 2009, and January 1 and 4, 2010. The segments highlighted in yellow, orange and magenta deserve particular attention since they are vegetated areas. A photo taken on the ground of the yellow object under non-flooded conditions is shown in the right panel of Fig. 16. The classification is obtained by using a fuzzy logic-based algorithm trained by the outputs of the simulations produced by applying the Tor

Fig. 16 Image obtained as result of the segmentation process applied to five COSMO-SkyMed images of the December 2009 Tuscany (Italy) flood (see text for the acquisition dates). The various segments are represented by different gray levels, except for some objects that are highlighted using different colors. A photo taken on the ground in the *yellow region* a few months after the flood is shown in the *right panel* (From Pulvirenti et al. [46], © 2012 IEEE)

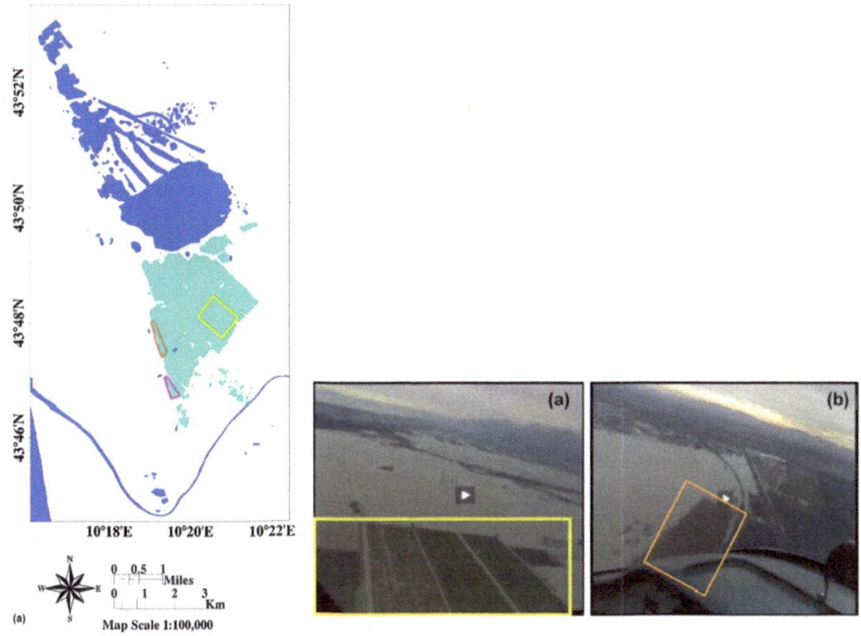

Fig. 17 *Left panel*: map of flooded areas produced using the COSMO-SkyMed image of the Tuscany (Italy) flood acquired on December 30, 2009. *White* non-flooded, *cyan* flooded, *blue* water bodies according to the CORINE Land Cover. *Central* and *right panels*: screenshots derived from a video found on the "you reporter" website (http://youreporter.it) during a flight performed on December 26, 2009 over the region hit by the inundation. The zones enclosed by the *yellow box* in the *central panel* and by the *orange box* in the *right panel* are highlighted with the same colors in the *left panel* (From Pulvirenti et al. [46], © 2012 IEEE)

Vergata model. The model is able to predict the increase of the backscatter due to the double-bounce effect caused by the presence of standing water. The result is shown in Fig. 17, together with two photos of the yellow and orange objects taken from an airplane after the flood occurrence. Even in this example, the possibility of SAR to detect floodwater beneath vegetation in some conditions (e.g., forests with low density of trees or leafless trees) clearly emerges. The double bounce represented the key to detect flooded vegetation also in this case study.

In the previous examples, the detection of the flooded vegetated and urban areas has been based on the detection of the double-bounce effect. This detection considered the change in the backscatter due to the change in the dielectric constant between soil and water as well as the decrease of the surface roughness (in the absence of significant winds). However, through this simplified modeling, not all the multitemporal radar signatures of flooded vegetation ad urban areas can be explained. An example of possible ambiguities in the radar signature of flooded vegetation is shown in Fig. 18, which refers to the COSMO-SkyMed images of the flood that hit the Alessandria district (northern Italy) in April 2009. The images

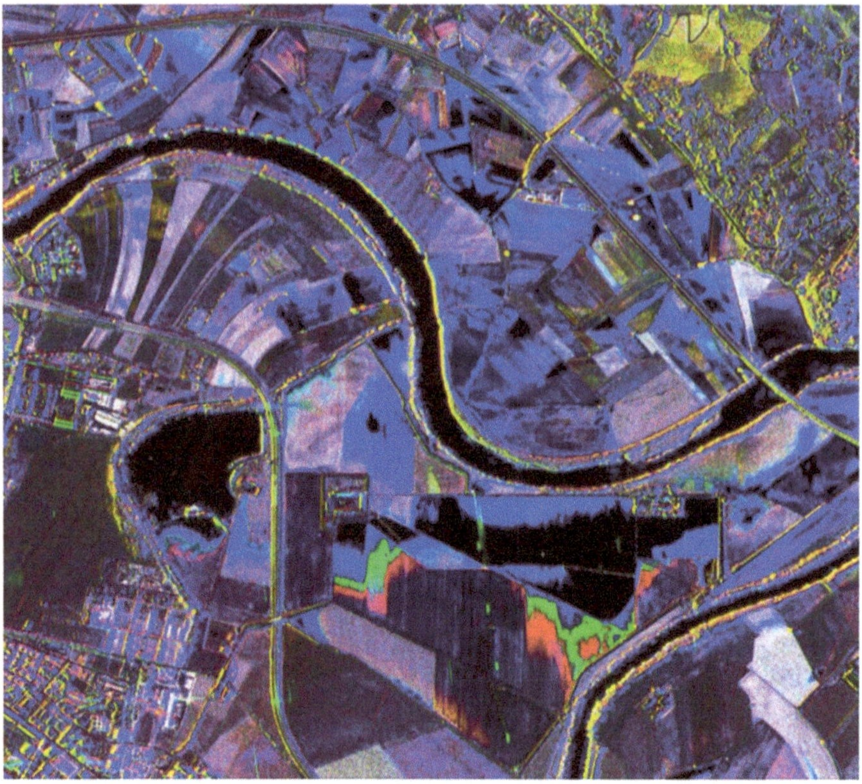

Fig. 18 RGB composition of three COSMO-SkyMed intensity images of the 2009 Alessandria (Italy) flood. *Red* April 29, 2009, *green* April 30, 2009, *blue* May 1, 2009 (From Pulvirenti et al. [45], © 2010 Elsevier Inc. All rights reserved)

were acquired on April 29, 2009, April 30, 2009, and May 1, 2009. In the zones that appear blue in Fig. 18, the radar return measured on May 1 is considerably larger than that measured on the two previous days. Some small areas appear red (high backscatter on April 29) and green (high backscatter on April 30). Interpreting these features is not a simple task, and again electromagnetic models turn out to be helpful in this respect. The water level was not considered in the previous examples. However, it may have a considerable role in determining the radar return from flooded crops (with relatively small height). In fact, an increase of the water level implies a decrease of the relative height of the stem emerging from water. This decrease may reduce the attenuation from the emerged vegetation but may also cause a decrease of the double-bounce effect (stems emerging from water are too short to produce a strong dihedral return). Even in flooded urban areas, the presence of standing water might increase the intensity of the double-bounce effect, but if the water level is not negligible with respect to the building height, the dihedral size diminishes, and a decrease of the radar cross section occurs [31].

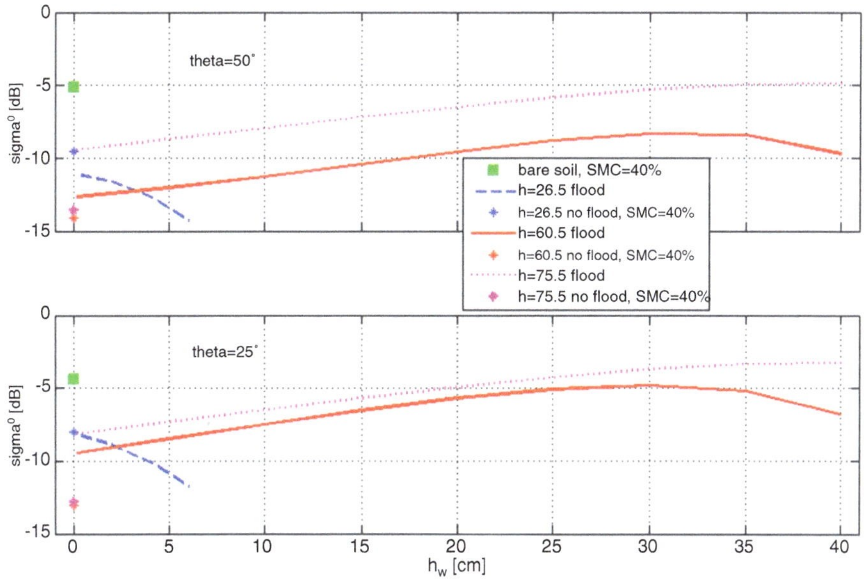

Fig. 19 Radar return versus water level (h_w) for a terrain covered by wheat, as predicted by the model used by the authors (X band, horizontal polarization). Three kinds of plant heights (h) are considered: $h = 26.5$ cm (*blue dashed line*), $h = 60.5$ cm (*red line*), and $h = 75.5$ cm (*magenta-dotted line*). The simulations performed under non-flooded conditions are also included, assuming SMC = 40%. *Upper panel*, incidence angle equal to 50°; *lower panel*, incidence angle equal to 25° (From Pulvirenti et al. [45], © 2010 Elsevier Inc. All rights reserved)

Figure 19 shows a simulation performed using the Tor Vergata model [7] that considers the water level as input parameter. It regards a small leaves' plant as wheat and X-band, horizontal polarization. By analyzing Fig. 19, it can be inferred that the effect of the entire wheat plant on the electromagnetic radiation is basically that of an attenuating layer. For non-flooded soils, the backscatter initially decreases with the growth of the plant and then remains approximately constant when scattering comes from the vegetation layer only (see asterisks and squares in the figure), i.e., the contribution of the soil is completely masked. For a wheat plant at an intermediate stage of growth ($h = 60.5$ cm; red solid lines), an increase of the water level implies that the scattering center of the double-bounce scattering mechanism moves toward the upper part of the stem, while the aforementioned attenuating vegetation layer becomes thinner. Consequently, the double-bounce backscattering becomes the predominant effect and causes an increase of the backscatter. For $h_w > 30$–35 cm, the emerging stem becomes so small that it is not capable of giving rise to a considerable double-bounce enhancement, and the backscatter decreases. As for $h = 26.5$ cm (blue dashed lines), the stem is short and thinner with respect to a developed plant, so that a further shortening due to the increase of h_w causes a decrease of the backscatter because the specular reflection due to the underlying water tends to become the predominant mechanism.

By accounting for the results of Fig. 19, it is possible to explain the multitemporal radar signatures of Fig. 18. In the areas that appear red in Fig. 18, there is a considerable decrease of the backscatter from April 29 to April 30, 2009. Such a large drop is caused by a change of the surface conditions from flooded to non-flooded. On April 29, the flood conditions give rise to a large double-bounce effect, and the drop of SAR intensity on the following day can be explained by assuming that on April 30 the receding of water is completed. In the areas that appear green in Fig. 18, there is a change from a specular-like surface to a situation in which a strong double-bounce effect occurs. Vegetation is completely covered by water on April 29, while, on April 30, the decrease of the water level implies that the emerging stems are tall enough to give rise to a strong double-bounce backscattering. As for the decay occurring from April 30 to May 1, the explanation is the same as that previously given for the red areas, that is the complete receding of the water. As for the blue areas, even though one can think that the increase of the SAR intensity from April 30 to May 1 can be due to the water receding, the presence of vegetation should be taken into account. This means that the aforementioned explanation is valid for bare fields but for vegetated soils is not always acceptable. In the latter case, an abrupt increase is again due to the transition from a situation of plants totally submerged by water to a situation in which the water level decreases and a strong double-bounce effect takes place. A flood evolution map based on the previous considerations is shown in Fig. 20.

4.4 *Urban Areas: Fuzzy Logic, Object-Based SAR Intensity and InSAR Coherence*

In the last examples discussed above, in order to tackle the ambiguities of the radar signature of flooded areas, we have relied on the analysis of multitemporal intensity data, possibly complemented by ancillary data, such as land cover maps or DEMs, and by the information provided by electromagnetic models. The detection of the double bounce was the key to remotely sense floodwater. However, the intensity of the double bounce contribution to the backscattered signal depends on several parameters. For instance, considering vegetation, single scattering and/or absorption due to canopy can mask the enhancement of the double-bounce mechanism. Even parameters as plant density and structure influence intensity of the double-bounce contribution. As for urban settlements, the double-bounce radar cross section decreases as the platform flight direction differs from the direction of alignment of the buildings and the adjacent roads [49]. As a consequence, the detection of the double bounce is not always possible using intensity data; in these cases the coherence may represent a source of complementary data.

An example of joint use of intensity and coherence data is shown in Fig. 21. It regards the COSMO-SkyMed images of the Japan tsunami occurred on March 11, 2011 and, in particular, an area of interest located in the Sendai harbor. The images

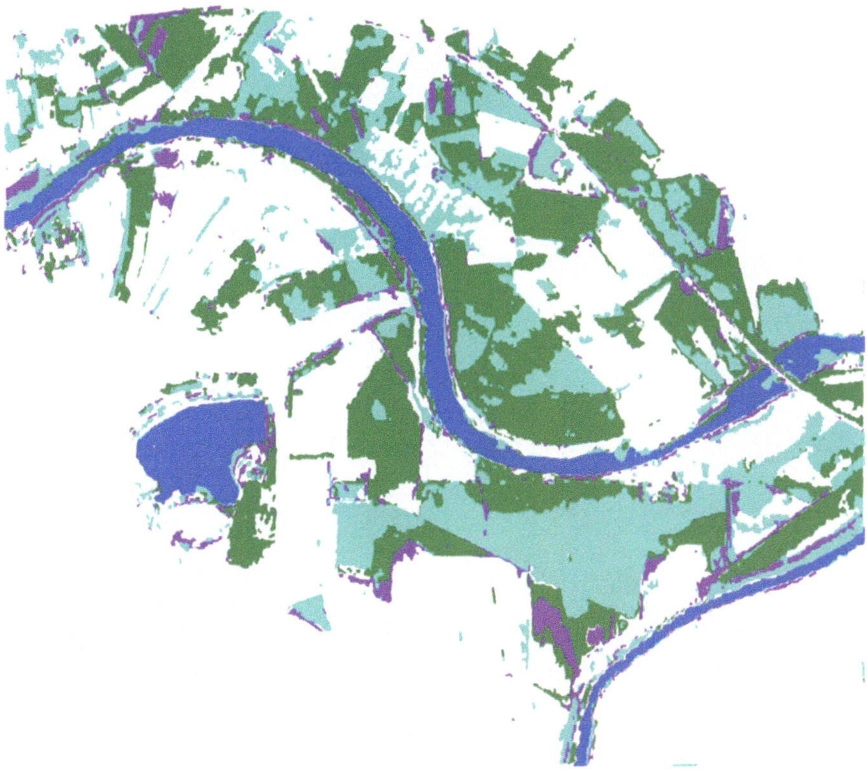

Fig. 20 Flood evolution map derived from the analysis of the COSMO-SkyMed images of the 2009 Alessandria (Italy) flood (see Fig. 18). *White* non-flooded, *sea green* flooded on April 29, *purple* flooded on April 29, *cyan* flooded throughout April 29–May 1, *blue* water body (From Pulvirenti et al. [45], © 2010 Elsevier Inc. All rights reserved)

were acquired 1 day and 2 days after the event in interferometric mode. It can be seen that there are some dark zones corresponding to sea and rivers characterized by low intensity and low coherence. In the blue area placed in the upper left part of the figure, the intensity increases on March 13; this means that in this area, which is prevalently bare, the flooding water present on March 12 recedes on March 13. For the Sendai case study, it is interesting to focus on the urban area in the central part of Fig. 21. There are again some blue pixels close to dark ones representing areas flooded on March 12 that get dry on the following day, and there are many pixels that appear cyan, i.e., low coherence between the images but high intensity in both the images (generally larger than 0 dB and reaching almost 7 dB; see [10]). For most of the pixels that appear cyan in Fig. 21, the intensity values do not change very much between March 12 and March 13, 2011. Low coherence between two images together with large backscattering that is similar between the two interferometric

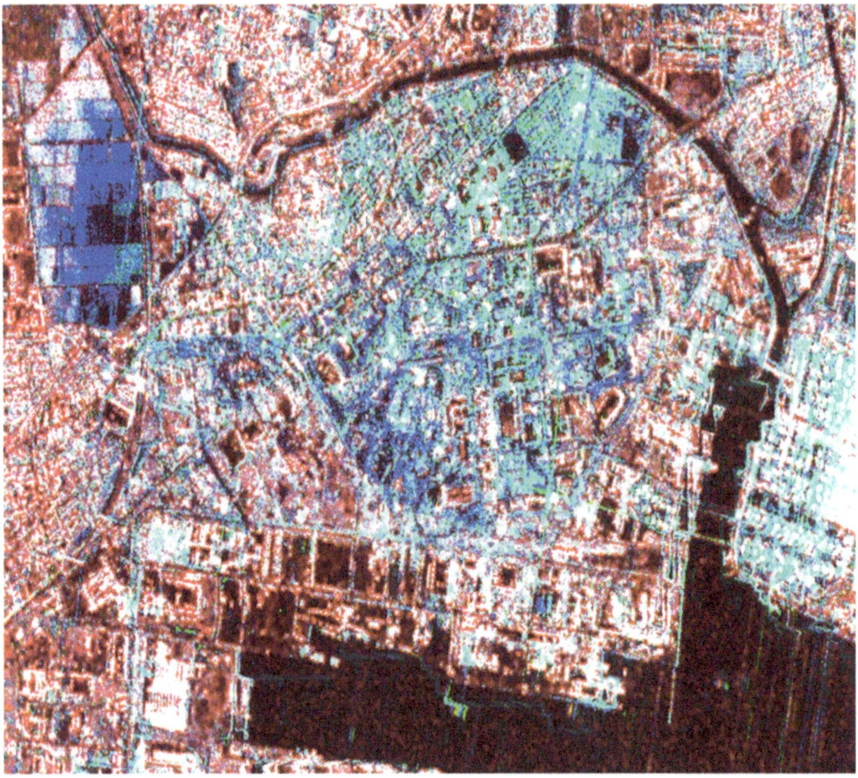

Fig. 21 RGB color composite images of the Sendai harbor created by using the COSMO-SkyMed observations of the 2011 Japan tsunami acquired on March 12 and March 13, 2011. *Red* coherence between the two images, *green* March 12 intensity, *blue* March 13 intensity (From Chini et al. [10], © 2012 IEEE)

acquisitions can be explained by supposing that cyan pixels represent flooded urban areas on both March 12 and March 13, 2011.

A second example regarding the joint use of intensity and coherence data refers to the flood that hit the Emilia–Romagna region (Italy) on January 19, 2014. Figure 22 shows the multitemporal intensity and coherence images created taking advantage of the COSMO-SkyMed interferometric acquisitions performed on December 16, 2013, January 5, 2014, and January 21, 22, and 23, 2014. The area hit by the flood is clearly visible; in particular, the area that appears in dark red tone in the left panel (low backscatter on January 21–22) represents the core of the flooded region. Few cyan areas denote an increase of the backscatter, while, for the magenta areas, the backscatter increase from January 21 to January 22, 2014 can be ascribed to the water receding. In the right panel of Fig. 22, the blue color prevails due to the small temporal baseline of the January 22–23 pair. The core of the flooded areas appears mainly in red tone (pre-event coherence higher than co-event and post-event ones)

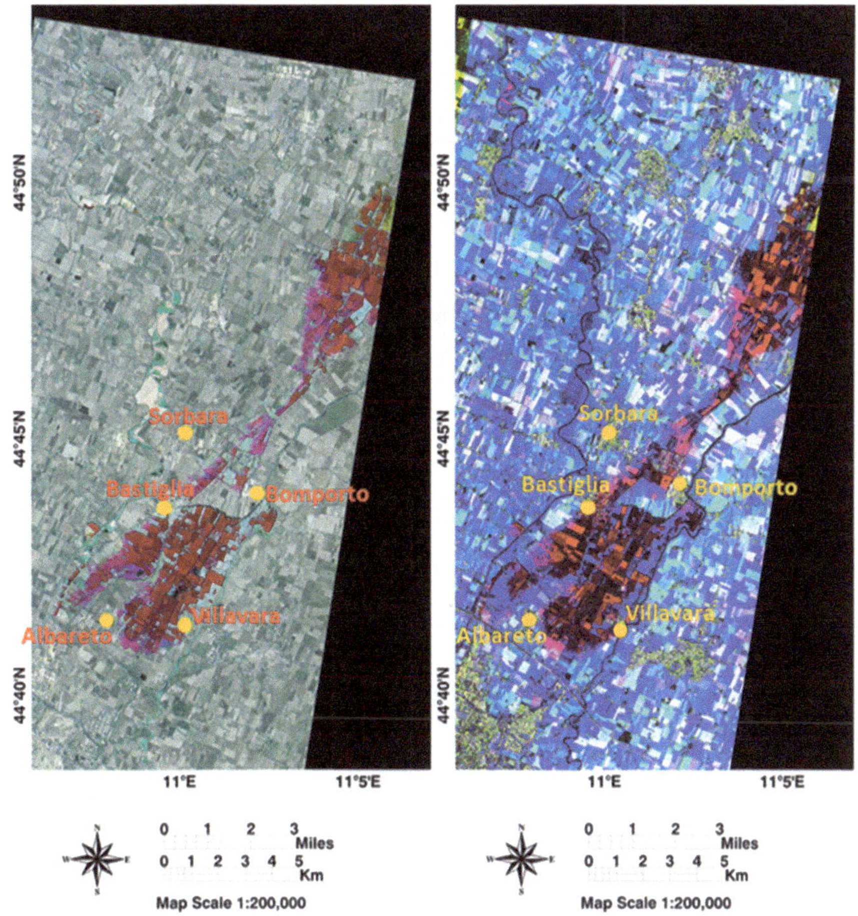

Fig. 22 *RGB* color composites of the area hit by the 2014 Emilia–Romagna flood created using COSMO-SkyMed data. *Left panel*: *R* January 5, 2014 intensity, *G* January 21, 2014 intensity, *B* January 22, 2014 intensity. *Right panel*: *R* pre-event coherence (December 16, 2013–January 5, 2014), *G* co-event coherence (January 5–21, 2014), *B* post-event coherence (January 22–23, 2014) (From Pulvirenti et al. [49], © 2015 IEEE)

and black tone (even the pre-event coherence is low because of the presence of vegetation or changes in soil roughness). For the topic of this chapter, the core of the flooded region (mainly corresponding to open water) is not very important, while it is interesting to focus on some vegetated and urban areas.

Figure 23 is the same as Fig. 22 but zoomed on a small area of interest. In particular, two regions of interest (ROI) are drawn and the multitemporal trends of the average values of intensity and coherence for these ROIs are shown in Fig. 24. Looking at Fig. 23, it can be seen that the ROIs have the same color in

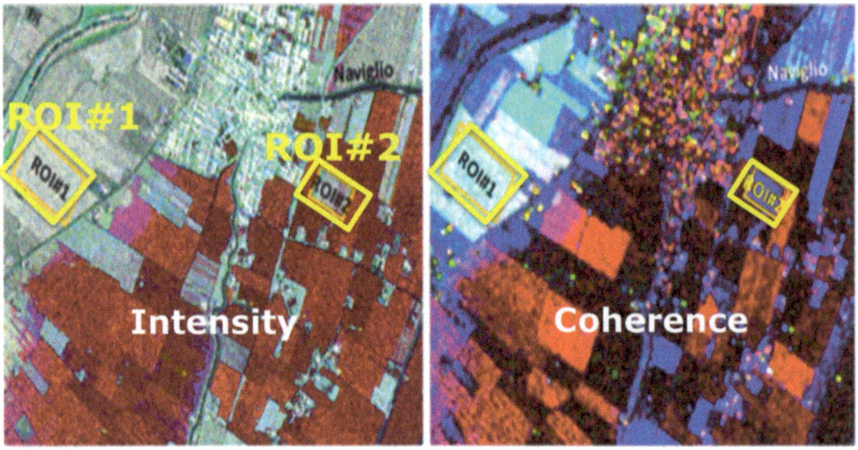

Fig. 23 Same as Fig. 22 but zoomed on an area of interest (From Pulvirenti et al. [49], © 2015 IEEE)

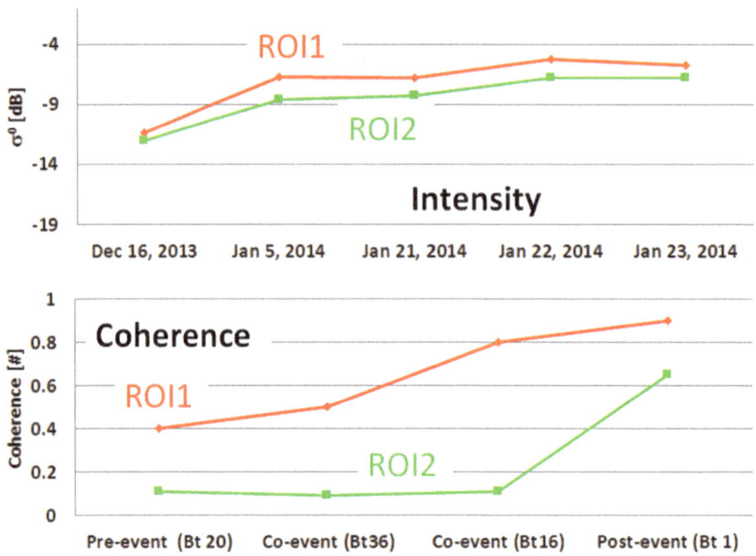

Fig. 24 Temporal trends of the average values of the backscatter (*upper panel*) and the coherence (*lower panel*) in the ROIs drawn in Fig. 23

the intensity RGB but different colors in the coherence RGB. This is reflected by the multitemporal trends that are similar for what concerns intensity and different for what concerns coherence. This means that using only the intensity data, these two areas would be likely classified as belonging to the same class (non-flooded areas). The pre-event coherence of ROI1 is considerably higher than that of ROI2, so that ROI1 was a bare or, at most, scarcely vegetated field, as opposed to ROI2.

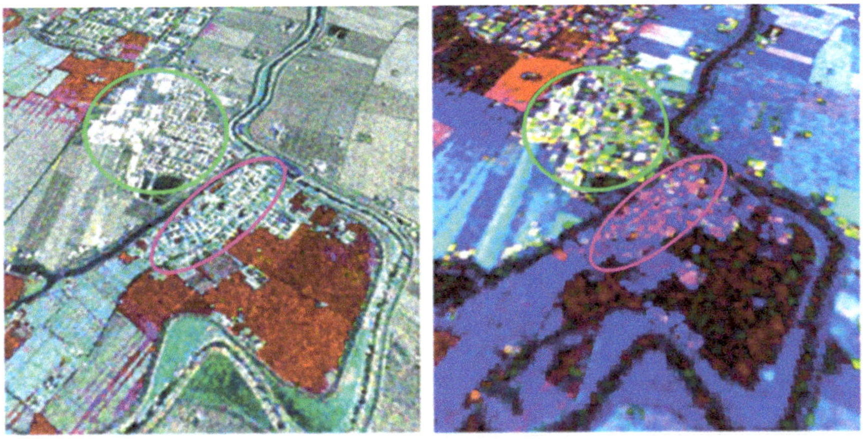

Fig. 25 Same as Fig. 22 but zoomed on the Bomporto village (From Pulvirenti et al. [49], © 2015 IEEE)

The temporal trend of the coherence for ROI1 supports the hypothesis that this field is not flooded; its increase is mostly related to the temporal baseline, and co-event coherences (December 16, 2013 to January 21, 2014 and January 5 to January 21, 2014) are high. For ROI2, the pre-event coherence is quite low, so that the field is vegetated (actually, it is a vineyard). The strong increase of the post-event coherence of ROI2 (from ~0.1 to ~0.7) indicates the occurrence of the double-bounce effect due to the presence of water. This double-bounce effect is not detected by intensity data (small difference between the backscatter of January 5 and January 21, 2014) probably because the contribution of the double-bounce (dihedral) effect to the total backscatter does not exceed the strong decrease of the soil backscatter due to the presence of standing water.

The 2014 Emilia–Romagna case study is very interesting for the purpose of mapping flooded urban areas. Figures 25 and 26 are analogous to Figs. 23 and 24, but for the urban area of Bomporto, in the Modena district. Two zones can be visually detected (highlighted by green and magenta ellipses). Even in this case, while the multitemporal trend of the SAR intensities is very similar, the trend of the coherence is totally different. The area highlighted by the green ellipse in Fig. 25 appears as white/gray, so that no substantial changes in the backscatter occurred in the period January 5–22. Conversely, in the area that appears in magenta tone in the right panel, the co-event coherence drops off due to the presence of floodwater on January 21, 2014. Even in this case, without the use of the coherence, this area would be labeled as non-flooded. Note that also for the urban area of Bomporto, the post-event coherence increases. As opposed to what happens for the post-event pair of images of the Japan tsunami, in this case the presence of water in urban areas implies an increase of the coherence. The reason can be ascribed to the fact that for the Japan test case, the water is not stagnant, and also a lot of debris is present

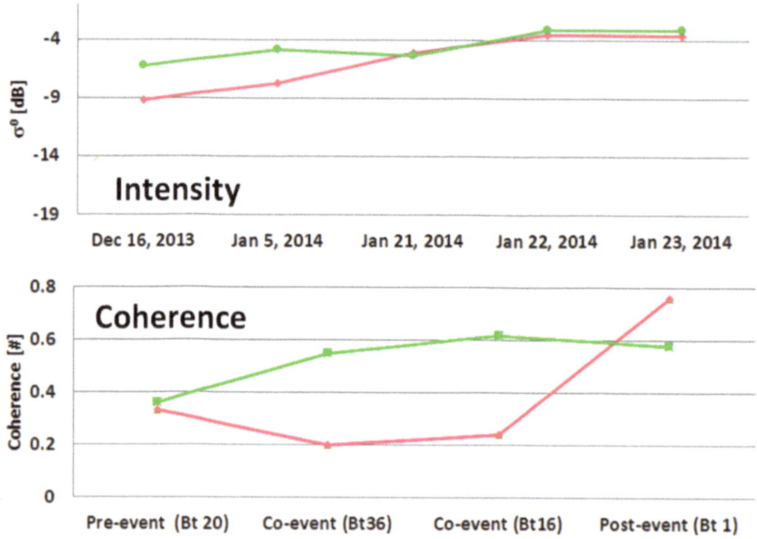

Fig. 26 Temporal trends of the average values of the backscatter (*left panel*) and the coherence (*right panel*) for the areas highlighted in *green* and *magenta* in Fig. 25 (Bomporto village)

[10], so that a decorrelation between post-event images occurs. For the Bomporto case study, on the contrary, stagnant water forms the horizontal surface of a double-bounce structure giving high coherence. Note that, again, intensity data are not able to detect the increase of the double bounce involving the smooth water surface and vertical walls.

The left panel of Fig. 27 shows the map of flooded areas obtained including in a classification algorithm based on a simple segmentation technique (region growing) both intensity and coherence data. It can be noted that the classification labels as flooded most of the urban pixels in the villages of Bastiglia and Villavara, whereas the town Bomporto results as partially flooded, in agreement with the previous discussion. The right panels of Fig. 27 show two photos taken in the Bomporto village and, in particular, in the zone highlighted by the magenta ellipse in Fig. 25 showing that this zone was actually flooded on January 22–23, 2014.

Acknowledgments The work of Marco Chini was supported by the National Research Fund of Luxembourg through the MOSQUITO (C15/SR/10380137) project. The COSMO-SkyMed © images are © of the Italian Space Agency (ASI), delivered by ASI under a license to use.

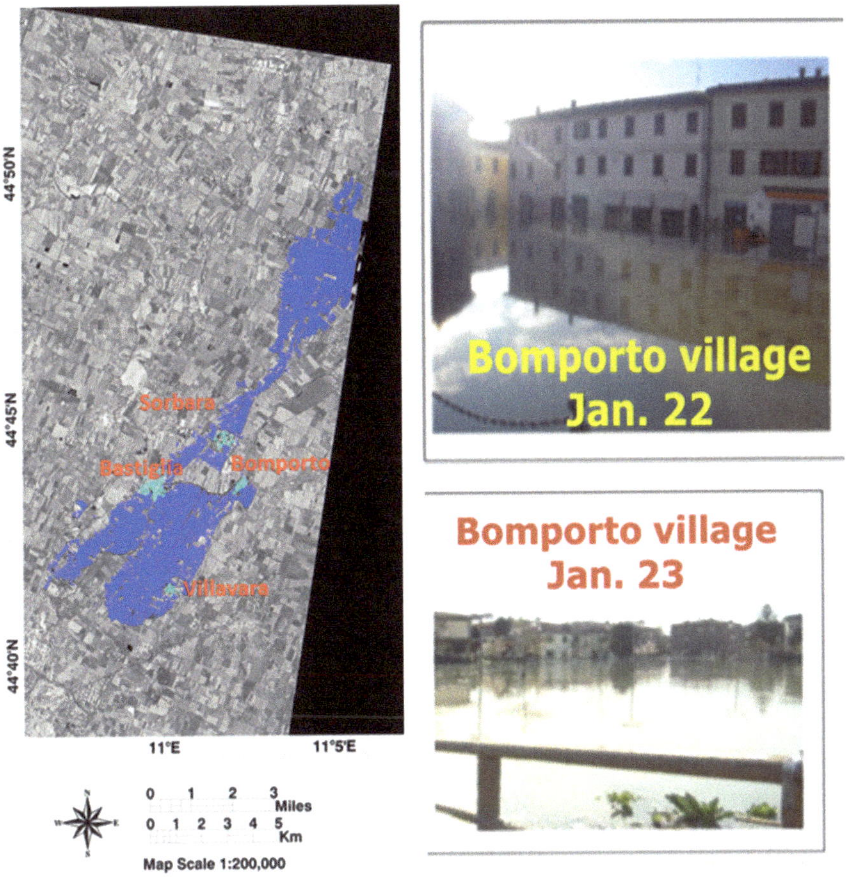

Fig. 27 *Left panel*: flood map for the 2014 Emilia–Romagna flood. *Blue* flooded agricultural areas, *cyan* flooded urban areas, background, January 21, 2014 intensity. *Right panels*: Photographs taken in the urban area of Bomporto (From Pulvirenti et al. [49], © 2015 IEEE)

References

1. Attema, E.P.W., Ulaby, F.T.: Vegetation modelled as a water cloud. Radio Sci. **13**, 357–364 (1978)
2. Barber, D.G., Hocheim, K.P., Dixon, R., Mosscrop, D.R., Mcmullan, M.J.: The role of earth observation technologies in flood mapping: A Manitoba case study. Can. J. Remote. Sens. **22**, 137–143 (1996)
3. Bazi, Y., Bruzzone, L., Melgani, F.: An unsupervised approach based on the generalized Gaussian model to automatic change detection in multitemporal SAR images. IEEE Trans. Geosci. Remote Sens. **43**(4), 874–887 (2005)
4. Benediktsson, J.A., Pesaresi, M., Arnason, K.: Classification and feature extraction for remote sensing images from urban areas based on morphological transformations. IEEE Trans. Geosci. Remote Sens. **41**, 1940–1949 (2003)

5. Bovolo, F., Bruzzone, L.: A split-based approach to unsupervised change detection in large-size multitemporal images: Application to tsunami-damage assessment. IEEE Trans. Geosci. Remote Sens. **45**(6), 1658–1670 (2007)
6. Bovolo, F., Bruzzone, L., Marconcini, M.: A novel approach to unsupervised change detection based on a semisupervised SVM and a similarity measure. IEEE Trans. Geosci. Remote Sens. **46**(7), 2070–2082 (2008)
7. Bracaglia, M., Ferrazzoli, P., Guerriero, L.: A fully polarimetric multiple scattering model for crops. Remote Sens. Environ. **54**, 170–179 (1995)
8. Chini, M., Bignami, C., Stramondo, S., Pierdicca, N.: Uplift and subsidence due to the December 26th, 2004, Indonesian earthquake and tsunami detected by SAR data. Int. J. Remote Sens. **29**(13), 3891–3910 (2008)
9. Chini, M., Pierdicca, N., Emery, W.J.: Exploiting SAR and VHR optical images to quantify damage caused by the 2003 bam earthquake. IEEE Trans. Geosci. Remote Sens. **47**(1), 145–152 (2009)
10. Chini, M., Pulvirenti, L., Pierdicca, N.: Analysis and interpretation of the COSMO-SkyMed observations of the 2011 Japan tsunami. IEEE Geosci. Remote Sens. Lett. **9**(3), 467–471 (2012)
11. Chini, M., Piscini, A., Cinti, F.R., Amici, S., Nappi, R., De Martini, P.M.: The 2011 Tohoku-Oki (Japan) tsunami inundation and liquefaction investigated by optical, thermal and SAR data. IEEE Geosci. Remote Sens. Lett. **10**(2), 347–351 (2013)
12. Chini, M., Albano, M., Saroli, M., Pulvirenti, L., Moro, M., Bignami, C., Falcucci, E., Gori, S., Modoni, G., Pierdicca, N., Stramondo, S.: Coseismic liquefaction phenomenon analysis by COSMO-SkyMed: 2012 Emilia (Italy) earthquake. Int. J. Appl. Earth Obs. Geoinf. **39**, 65–78 (2015)
13. Chini, M.: Building damage from multi-resolution, object-based, classification techniques. In: Beer, M., Kougioumtzoglou, I.A., Patelli, E., Au, I.S.-K. (eds.) Encyclopaedia of Earthquake Engineering, pp. 350–360. Springer, Berlin (2015). doi:10.1007/978–3–642-35344-4
14. Chini, M., Papastergios, A., Pulvirenti, L., Pierdicca, N., Matgen, P., Parcharidis, I.: SAR coherence and polarimetric information for improving flood mapping, Proceeding IEEE Igarss 2016, Beijing (China), pp. 7577–7580, (2016)
15. Chini, M., Hostache, R., Giustarini, L., Matgen, P.: A hierarchical split-based approach for parametric thresholding of SAR images: Flood inundation as a test case. IEEE Trans. Geosci. Remote Sens. 1–14 (2017). https://doi.org/10.1109/TGRS.2017.2737664
16. Cloude, S.R., Pottier, E.: An entropy based classification scheme for land applications of polarimetric SAR. IEEE Trans. Geosci. Remote Sens. **35**, 68–78 (1997)
17. D'Addabbo, A., Refice, A., Pasquariello, G., Lovergine, F.P., Capolongo, D., Manfreda, S.: A Bayesian network for flood detection combining SAR imagery and ancillary data. IEEE Trans. Geosci. Remote Sens. **54**(6.), 7429750), 3612–3625 (2016)
18. Ferrazzoli, P., Guerriero, L.: Radar sensitivity to tree geometry and woody volume: A model analysis. IEEE Trans. Geosci. Remote Sens. **33**(2), 360–371 (1995)
19. Ferro, A., Brunner, D., Bruzzone, L., Lemoine, G.: On the relationship between double bounce and the orientation of buildings in VHR SAR images. IEEE Geosci. Remote Sens. Lett. **8**(4), 612–616 (2011)
20. Franceschetti, G., Iodice, A., Riccio, D.: A canonical problem in electromagnetic backscattering from buildings. IEEE Trans. Geosci. Remote Sens. **40**(8), 1787–1801 (2002)
21. Freeman, A., Durden, S.L.: A three-component scattering model for polarimetric SAR data. IEEE Trans. Geosci. Remote Sens. **36**, 963–973 (1998)
22. Giustarini, L., Matgen, P., Hostache, R., Plaza, D., Pauwels, V.R.N., Lannoy, G.J., Keyser, R.D., Pfister, L., Hoffmann, L., Savenije, H.H.G.: Assimilating SAR-derived water level data into a flood model: A case study. Hydrol. Earth Syst. Sci. **15**(7), 2349–2365 (2011)
23. Giustarini, L., Hostache, R., Matgen, P., Schumann, G.J.-P., Bates, P.D., Mason, D.C.: A change detection approach to flood mapping in urban areas using TerraSAR-X. IEEE Trans. Geosci. Remote Sens. **51**(4), 2417–2430 (2013)

24. Giustarini, L., Vernieuwe, H., Verwaeren, J., Chini, M., Hostache, R., Matgen, P., Verhoest, N.E.C., De Baets, B.: Accounting for image uncertainty in SAR-based flood mapping. Int. J. Appl. Earth Obs. Geoinf. **34**, 70–77 (2015)
25. Giustarini, L., Chini, M., Hostache, R., Pappenberger, F., Matgen, P.: Flood hazard mapping combining hydrodynamic modelling and multi annual remote sensing data. Remote Sens. **7**(10), 14200–14226 (2015)
26. Giustarini, L., Hostache, R., Kavetski, D., Chini, M., Corato, G., Schlaffer, S., Matgen, P.: Probabilistic flood mapping using synthetic aperture radar data. IEEE Trans. Geosci. Remote Sens. **54**, 6958–6969 (2016)
27. Henry, J.-B., Chastanet, P., Fellah, K., Desnos, Y.-L.: Envisat multi-polarized ASAR data for flood mapping. Int. J. Remote Sens. **27**(10), 1921–1929 (2006)
28. Hess, L.L., Melack, J.M., Simonett, D.S.: Radar detection of flooding beneath the forest canopy: A review. Int. J. Remote Sens. **11**, 1313–1325 (1990)
29. Hess, L.L., Melack, J.M., Novo, E.M.L.M., Barbosa, C.C.F., Gastil, M.: Dual-season mapping of wetland inundation and vegetation for the central Amazon basin. Remote Sens. Environ. **87**, 404–428 (2003)
30. Hostache, R., Matgen, P., Schumann, G., Puech, C., Hoffmann, L., Pfister, L.: Water level estimation and reduction of hydraulic model calibration uncertainties using satellite SAR images of floods. IEEE Trans. Geosci. Remote Sens. **47**(2), 431–441 (2009)
31. Iervolino, P., Guida, R., Iodice, A., Riccio, D.: Flooding water depth estimation with high-resolution SAR. IEEE. **53**, 2295–2307 (2015)
32. Kim, S.B., Ouellette, J.D., van Zyl, J.J., Johnson, J.T.: Detection of inland open water surfaces using dual polarization L-band radar for the soil moisture active passive mission. IEEE Trans. Geosci. Remote Sens. **54**(6), 3388–3399 (2016)
33. Martinis, S., Twele, A., Voigt, S.: Towards operational near real-time flood detection using a split-based automatic thresholding procedure on high resolution TerraSAR-X data. Nat. Hazards Earth Syst. Sci. **9**, 303–314 (2009)
34. Martinis, S., Kersten, J., Twele, A.: A fully automated TerraSAR-X based flood service. ISPRS J. Photogramm. Remote Sens. **104**, 203–212 (2015)
35. Mason, D.C., Giustarini, L., Garcia-Pintado, J., Cloke, H.L.: Detection of flooded urban areas in high resolution Synthetic Aperture Radar images using double scattering. Int. J. Appl. Earth Obs. Geoinf. **28**, 150–159 (2014)
36. Matgen, P., Hostache, R., Schumann, G., Pfister, L., Hoffmann, L., Savenije, H.H.G.: Towards an automated SAR-based flood monitoring system: Lessons learned from two case studies. Phys. Chem. Earth. **36**, 241–252 (2011)
37. Pal, S.K., Rosenfeld, A.: Image enhancement and thresholding by optimization of fuzzy compactness. Pattern Recogn. Lett. **7**, 77–86 (1988)
38. Panegrossi, G., Ferretti, R., Pulvirenti, L., Pierdicca, N.: Impact of ASAR soil moisture data on the MM5 precipitation forecast for the Tanaro flood event of April 2009. Nat. Hazard. Earth Syst. Sci. **11**, 3135–3149 (2011)
39. Pesaresi, M., Benediktsson, J.A.: A new approach for the morphological segmentation of high-resolution satellite imagery. IEEE Trans. Geosci. Remote Sens. **39**, 309–320 (2001)
40. Pierdicca, N., Castracane, P., Pulvirenti, L.: Inversion of electromagnetic models for bare soil parameter estimation from multifrequency polarimetric SAR data. Sensors. **8**, 8181–8200 (2008)
41. Pierdicca, N., Chini, M., Pulvirenti, L., Macina, F.: Integrating physical and topographic information into a fuzzy scheme to map flooded area by SAR. Sensors. **8**, 4151–4164 (2008)
42. Pierdicca, N., Pulvirenti, L.: Comparing scatterometric and radiometric simulations with geophysical model functions to tune a sea wave spectrum model. IEEE Trans. Geosci. Remote Sens. **46**, 3756–3767 (2008)
43. Pierdicca, N., Pulvirenti, L., Chini, M., Guerriero, L., Candela, L.: Observing floods from space: Experience gained from COSMO-SkyMed observations. Acta Astron. **84**, 122–133 (2013)

44. Pulvirenti, L., Pierdicca, N., Chini, M., Guerriero, L.: An algorithm for operational flood mapping from synthetic aperture radar (SAR) data based on the fuzzy logic. Nat. Hazard. Earth Syst. Sci. **11**(2), 529–540 (2011)
45. Pulvirenti, L., Chini, M., Pierdicca, N., Guerriero, L., Ferrazzoli, P.: Flood monitoring using multi-temporal COSMO-SkyMed data: Image segmentation and signature interpretation. Remote Sens. Environ. **115**(4), 990–1002 (2011)
46. Pulvirenti, L., Pierdicca, N., Chini, M., Guerriero, L.: Monitoring flood evolution in agricultural areas using COSMO-SkyMed data: The Tuscany 2009 case study. IEEE J. Select. Topics Appl. Earth Obs. Remote Sens. **6**(4), 1199–1210 (2013)
47. Pulvirenti, L., Pierdicca, N., Boni, G., Fiorini, M., Rudari, R.: Flood damage assessment through multitemporal COSMO-SkyMed data and hydrodynamic models: The Albania 2010 case study. IEEE J. Sel. Topics Appl. Earth Observ. Remote Sens. **7**(7), 2848–2855 (2014)
48. Pulvirenti, L., Marzano, F.S., Pierdicca, N., Mori, S., Chini, M.: Discrimination of water surfaces, heavy rainfall and wet snow using COSMO-SkyMed observations of severe weather events. IEEE Trans. Geosci. Remote Sens. **52**(2), 858–869 (2014)
49. Pulvirenti, L., Chini, M., Pierdicca, N., Boni, G.: Use of SAR data for detecting floodwater in urban and agricultural areas: The role of the interferometric coherence. IEEE Trans. Geosci. Remote Sens. **54**(3), 1532–1544 (2016)
50. Quartulli, M., Datcu, M.: Stochastic geometrical modeling for built-up area understanding from a single SAR intensity image with meter resolution. IEEE Trans. Geosci. Remote Sens. **42**, 1996–2003 (2004)
51. Schlaffer, S., Matgen, P., Hollaus, M., Wagner, W.: Flood detection from multi-temporal SAR data using harmonic analysis and change detection. Int. J. Appl. Earth Obs. Geoinform. **38**, 15–24 (2015)
52. Schlaffer, S., Chini, M., Dettmering, D., Wagner, W.: Mapping Wetlands in Zambia using seasonal backscatter signatures derived from ENVISAT ASAR time series. Remote Sens. **8**, 1–24 (2016)
53. Schlaffer, S., Chini, M., Giustarini, L., Matgen, P.: Probabilistic mapping of flood-induced backscatter changes in SAR time series. Int. J. Appl. Earth Obs. Geoinf. **56**, 77–87 (2017)
54. Schumann, G.J.-P., Frye, S., Wells, G., Adler, R., Brakenridge, R., Bolten, J., Murray, J., Slayback, D., Policelli, F., Kirschbaum, D., Wu, H., Cappelaere, P., Howard, T., Flamig, Z., Clark, R., Stough, T., Chini, M., Matgen, P., Green, D., Jones, B.: Unlocking the full potential of earth observation during the 2015 Texas flood disaster. Water Resour. Res. **52**, 3288–3293 (2016)
55. Shimada, T., Kawamura, H., Shimada, M.: An L-band geophysical model function for SAR wind retrieval using JERS-1 SAR. IEEE Trans. Geosci. Remote Sens. **41**, 518–531 (2003)
56. Soille, P.: Morphological Image Analysis – Principles and Applications. Springer, Berlin (2003)
57. Stramondo, S., Bignami, C., Chini, M., Pierdicca, N., Tertulliani, A.: Satellite radar and optical remote sensing for earthquake damage detection: Results from different case studies. Int. J. Remote Sens. **27**(20), 4433–4447 (2006)
58. Thiele, Cadario, E., Schulz, K., Thönnessen, U., Soerge, U.: Building recognition from multi-aspect high-resolution InSAR data in urban areas. IEEE Trans. Geosci. Remote Sens. **45**(11), 3583–3593 (2007)
59. Ticconi, F., Pulvirenti, L., Pierdicca, N.: Models for scattering from rough surfaces. In: Zhurbenko, V. (ed.) Electromagnetic Waves, Croatia. In Tech. ISBN: 978–953–307–304–0 (2011)
60. Ulaby, F.T., Sarabandi, K., McDonald, K., Whitt, M., Dobson, M.C.: Michigan microwave canopy scattering model. Int. Remote Sens. **11**(7), 1223–1253 (1990)
61. Wang, Y., Hess, L.L., Filoso, S., Melack, J.M.: Understanding the radar backscattering from flooded and nonflooded Amazonian forests: Results from canopy backscatter modeling. Remote Sens. Environ. **54**, 324–332 (1995)
62. Weydahl, D.J.: Flood monitoring in Norway using ERS-1 SAR images. In: Geoscience and Remote Sensing Symposium, 1996. IGARSS '96, Vol.1, pp. 151–153. Lincoln (1996)

63. Wood, M., Hostache, R., Neal, J., Wagener, T., Giustarini, L., Chini, M., Corato, G., Matgen, P., Bates, P.: Calibration of channel depth and friction parameters in the LISFLOOD-FP hydraulic model using medium resolution SAR data. Hydrol. Earth Syst. Sci. **20**(12), 4983–4997 (2016)
64. Zadeh, L.A.: Fuzzy sets. Inf. Control. **8**, 338–353 (1965)
65. Zebker, H.A., Villasenor, J.: Decorrelation in interferometric radar echoes. IEEE Trans. Geosci. Remote Sens. **30**(5), 950–959 (1992)

Data Fusion Through Bayesian Methods for Flood Monitoring from Remotely Sensed Data

Annarita D'Addabbo, Alberto Refice, Domenico Capolongo, Guido Pasquariello, and Salvatore Manfreda

Abstract Producing high-precision flood maps requires integrating and correctly classifying information coming from heterogeneous sources. Methods to perform such integration have to rely on different knowledge bases. A useful tool to perform this task consists in the use of Bayesian methods to assign probabilities to areas being subject to flood phenomena, fusing a priori information and modeling with data coming from radar or optical imagery. In this chapter we review the use of Bayesian networks, an elegant framework to cast probabilistic descriptions of complex systems, applied to flood monitoring from multi-sensor, multi-temporal remotely sensed and ancillary data.

Keywords Data fusion • Bayesian networks • Change detection • Time series analysis

1 Introduction

Flood monitoring from remotely sensed images can be considered as a particular instance of the more general environmental change detection issue. To capture any environmental change, multi-temporal images of the same area have to be managed. Moreover, some environment types are so complex and challenging that it is optimistic to assume that a single sensor can provide all the information required for their characterization [13]. In addition, satellite data have the great

---

A. D'Addabbo (✉) • A. Refice • G. Pasquariello
ISSIA-CNR, Bari, Italy
e-mail: daddabbo@ba.issia.cnr.it; refice@ba.issia.cnr.it; pasquariello@ba.issia.cnr.it

D. Capolongo
Department of Earth and Environmental Sciences, University of Bari, Bari, Italy
e-mail: domenico.capolongo@uniba.it

S. Manfreda
Department of European and Mediterranean Cultures: Architecture, Environment, Cultural Heritage (DiCEM), University of Basilicata, Potenza, Italy
e-mail: salvatore.manfreda@unibas.it

A. Refice et al. (eds.), *Flood Monitoring through Remote Sensing*,
Springer Remote Sensing/Photogrammetry,
https://doi.org/10.1007/978-3-319-63959-8_8

advantage of giving a snapshot of a very large area in a short time, but other information may be needed to get precise knowledge about some critical areas. Natural disaster scenarios are typical examples of complex situations in which different factors have to be considered to provide accurate and robust assessments. In particular, in flood detection, the presence of many land cover types, each one with a particular signature in the presence of inundations, requires modeling the behavior of different "objects" in remotely sensed images (not necessarily corresponding to land cover categories) in order to associate them to flood or no-flood conditions. Generally, the fusion of data, such as multi-temporal, multi-sensor, multi-resolution, and multi-platform image data, together with other information provided by various sources, seems essential for a consistent interpretation of complex scenes: the information obtained by data synergy exploitation is better (qualitatively and quantitatively, in terms of accuracy, robustness, etc.) than the one obtained from each source used individually [9]. In the case of flooding, information provided by human expertise or objective measurements such as distance from the river, terrain elevation, hydrologic information, or some combination thereof, can be opportunely merged with remotely sensed images to lead to better performances in detection. So, effective methods able to manage different types of data and images and exploit their information are needed.

In this chapter, we describe a class of statistical methods for estimating flood extension by the fusion of remotely sensed images, acquired from different sensors at different times, and ancillary data. The proposed method is based on probabilistic graphical models, an elegant framework which combines uncertainty (probabilities) and logical structure (independence constraints) to compactly represent complex, real-world phenomena [14, Ch. 2]. In particular, it is based on Bayesian networks (BNs), one of the most common types of graphical models, originated from the classical Bayesian inference theory [1, 18]. In recent years, BNs and, more generally, probabilistic graphical models have been used as data mining tools in many information fields, but they have rarely been used to process remote sensing data. In [27], a BN has been applied to two Landsat TM data acquired at different times on the same area, obtaining an overall change detection classification accuracy that overcomes the one obtained by other methods. However, to our knowledge, the capability of BNs to join different kinds of data has been scarcely exploited. In [12], a simple probabilistic graphical model has been applied to estimate the state of roads during flooding, combining a digital elevation model (DEM) with satellite optical images: the obtained results outperform other methods based only on the exploitation of imagery data or only on DEM information. On the other hand, BNs have been successfully used to construct risk assessment systems for natural disasters, such as rockfalls [33], avalanches [16], or wildfires [23]: various kinds of data, such as land cover types extracted from the Corine Land Cover database and road and building densities, have been fused by means of BNs to obtain reliable risk maps, although none included remotely sensed imagery. Here, BNs are presented and critically analyzed in a flood detection problem. By using some test cases, we show how a BN, opportunely conceived and designed, can be used to integrate multi-temporal remote sensed images and ancillary data, allowing to both mitigate the false alarms and to correctly identify flooded areas in events characterized by complex land cover ground conditions and time evolution. The final goal is the implementation of an automatic tool able to provide, as output products, probabilistic flood maps, i.e., maps having, for each pixel, the probability value that the corresponding area has been reached by the inundation, by considering as input data:

- M multi-temporal imagery sets, each one composed of $m_i; i = 1, \ldots, M$ images acquired at different times, on the same geographical area from the same ith source. Various sources on different platforms (such as satellite radar and optical data, aerial photography, imagery acquired by camera on autonomous vehicles, etc.) can be considered.
- N ancillary information data sets from different sources (such as digital elevation models, expert information, geological and hydrogeological models, land use/cover maps, etc.).

This chapter is organized as follows. In Sect. 2, the theoretical principles at the basis of the proposed method are discussed in details. Some experimental applications are presented and critically analyzed in Sect. 3, together with information on the study area and the considered flood event, the available data, as well as the reference data used for the evaluation. Finally, in Sect. 4, some conclusions are drawn.

2 Bayesian Networks for Flood Map Production

2.1 BN Basic Principles

BNs combine probability theory and graph theory, introducing graph structures into a probabilistic model to represent dependence assumptions among the involved variables [1, 18]. A BN is characterized by many attractive properties that can be particularly useful in modeling many real-world systems. First of all, it naturally provides a general methodology for handling data fusion, allowing the joint distribution of all variables to be written down tractably, even in cases where its explicit representation is particularly large. Moreover, a BN graphical structure can be based on prior knowledge, allowing to construct models that provide an accurate reflection of the understanding of a domain. Alternatively, the structure may be learned from data, or even a combination of the two approaches can be used: pose a prior BN and update it based on observations. Finally, it enables inference: as previously mentioned, the BN captures the joint distribution of all variables, and via inference, any/all conditional distribution(s) of interest, in any direction, can be expressed. These features appear particularly suited for the application domain of natural disaster monitoring and assessment [35]. The interactions of the variables of a "system" are encoded through a directed acyclic graph (DAG; see Fig. 1): the nodes of the DAG represent the random variables, and the edges linking them correspond to direct influence of one node on another, modeled as conditional

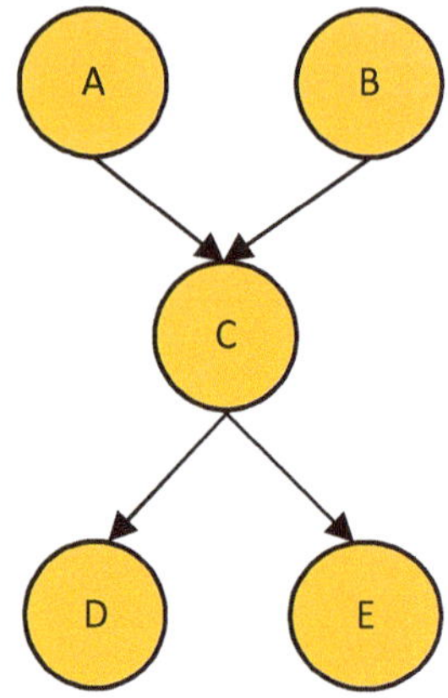

Fig. 1 DAG example

probabilities [1, 18]. In a graph, for each i and j, if $X_i \rightarrow X_j$, the following definitions hold:

- The variable X_i is a parent of the variable X_j.
- The variable X_j is a descendant of the variable X_j.

For example, in the graph depicted in Fig. 1, the variable C is a parent of D and E and it is a descendant of A and B.

The BN structure thus provides information about the underlying processes and the way various variables communicate and share "information" as it spreads around the network. Formally speaking, the DAG encodes conditional independence assumptions between variables, and the following assumption holds: each random variable X is conditionally independent from its non-descendants, i.e., it is independent from its non-descendants given its parents. For example, in the DAG in Fig. 1, the variables D and E are conditionally independent given their parent C, i.e., $p(D, E|C) = p(D|C) \cdot p(E|C)$. By considering this assumption, the DAG in Fig. 1 represents the following joint probability between the involved variables:

$$P(A, B, C, D, E) = P(A) \cdot P(B) \cdot P(C|A, B) \cdot P(D|C) \cdot P(E|C) \tag{1}$$

Further insight about this formula will be given in the following subsection, when the particular application to the flood detection problem will be faced.

2.2 *BN Structure for Flood Detection and Involved Variables*

In the flood detection problem, the BN framework helps to elegantly gather together several approaches which combine imagery data, external a priori information extracted from simple data such as distance from river courses, or more complex ones such as hydrologic flood models, or, e.g., terrain height or slope. Moreover,

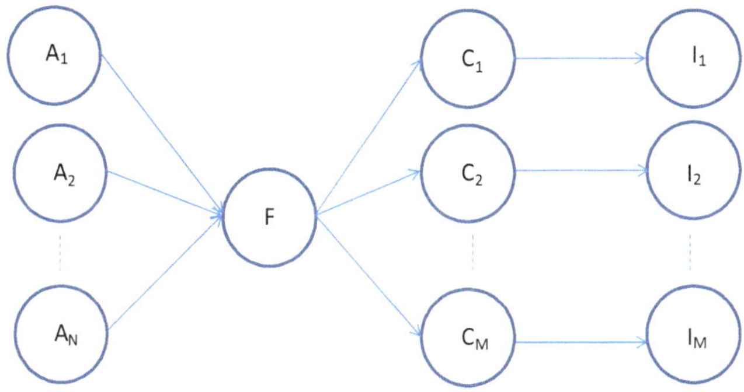

Fig. 2 BN scheme

it is also able to accommodate a description of a part of the temporal dynamics of the flooding event, by highlighting information from multi-temporal acquisitions. A general BN structure is considered and analyzed, by varying the number of imagery sources and ancillary data types and by considering different ways to merge their information content. The corresponding DAG is reported in Fig. 2.

A detailed description of the random variables considered is given below:

- The variable F is discrete and consists of only two states, flood and no-flood, so that $P(F = \text{flood}) = 1 - P(F = \text{noflood})$. This is the variable that we want to estimate by statistical inference.
- Each variable $I_i; i = 1, \ldots, M$ corresponds to a m_i-dimensional vector obtained from the imagery. It is worth noting that this general scheme encompasses different ways to merge the imagery data: one obvious possibility is to consider the M variables $I_1, I_2, \ldots, I_M$ as corresponding, respectively, to $m_1, m_2, \ldots, m_M$-dimensional vectors obtained from M different multi-temporal sets of images, each one acquired by a different source; at the other extreme, all available images, even acquired by different sensors and/or at different times, may be considered as corresponding to only one node, i.e., to a single, multi-dimensional variable, retaining internal dependencies among the various sources, and analyzed together.
- Since often the flood state does not exhibit a simple "causal" relation with the imagery, the intermediate variables $C_i; i = 1, \ldots, M$, are introduced [12]. Each C_i variable consists of K_i possible states, $C_{i,j}; j = 1, \ldots, K_i$, that can be determined either by using some a priori knowledge on the scene or by extracting them automatically from the imagery. Anyway, the number K_i of possible states and the usefulness of each C_i variable is strongly dependent on the complexity of the analyzed scene. For example, if the area under observation is characterized only by unvegetated/barren areas, and a single SAR intensity image is available, the variable C is unnecessary, and the I random variable can be directly connected to F: the backscatter values of flooded areas, in this case, will be very low and well separated from nonflooded ones. Otherwise, for more complex scenarios,

such as in presence of vegetated, forested, or urban areas, or if we want to follow the flood dynamics thanks to the availability of multi-temporal data, the introduction of the intermediate variable C_i allows to classify the various scattering behaviors into a finite number of types and then to focus the attention on the most interesting ones [12, 26].

- The variables $A_k; k = 1, \ldots, N$ represent some ancillary information, coming from different sources, not pertaining to remote sensing data.

2.3 Flood Probability Model Derived from the BN

In order to obtain flood probability maps, we are interested to infer the following conditional probability value

$$p(F = \text{flood}|I_1 = I_1^*, \ldots, I_M = I_M^*, A_1 = A_1^*, \ldots, A_N = A_N^*), \tag{2}$$

where $I_i^*, i = 1, \ldots, M$, and $A_k^*, k = 1, \ldots, N$ are the observed values for the random variables $I_i, i = 1, \ldots, M$, and $A_k, k = 1, \ldots, N$, respectively. This conditional probability can be derived from the joint probability $p(F, I_1, \ldots, I_M, C_1, \ldots, C_M, A_1, \ldots, A_N)$ related to the graph in Fig. 2, because the following relations hold:

$$\begin{aligned}
&p(F = \text{flood}|I_1 = I_1^*, \ldots, I_M = I_M^*, A_1 = A_1^*, \ldots, A_N = A_N^*) \\
&= \frac{p(F=\text{flood}, I_1=I_1^*, \ldots, I_M=I_M^*, A_1=A_1^*, \ldots, A_N=A_N^*)}{p(I_1=I_1^*, \ldots, I_M=I_M^*, A_1=A_1^*, \ldots, A_N=A_N^*)} \\
&= \frac{\sum_{C_1} \cdots \sum_{C_M} p(F=\text{flood}, I_1=I_1^*, \ldots, I_M=I_M^*, C_1, \ldots, C_M, A_1=A_1^*, \ldots, A_N=A_N^*)}{\sum_{C_1} \cdots \sum_{C_M} p(I_1=I_1^*, \ldots, I_M=I_M^*, C_1, \ldots, C_M, A_1=A_1^*, \ldots, A_N=A_N^*)} \\
&= \frac{\sum_{C_1} \cdots \sum_{C_M} p(F=\text{flood}, I_1=I_1^*, \ldots, I_M=I_M^*, C_1, \ldots, C_M, A_1=A_1^*, \ldots, A_N=A_N^*)}{\sum_F \sum_{C_1} \cdots \sum_{C_M} p(F, I_1=I_1^*, \ldots, I_M=I_M^*, C_1, \ldots, C_M, A_1=A_1^*, \ldots, A_N=A_N^*)}.
\end{aligned} \tag{3}$$

The probability $p(F, I_1, \ldots, I_M, C_1, \ldots, C_M, A_1, \ldots, A_N)$ in (3) can be easily factorized by considering the following conditional independence assumptions that hold between variables:

- Each I_i term, with $i = 1, \ldots, M$, is conditionally independent from all other nodes in the network given its parent C_i. In other words, if we know the state of the class C_i, our belief about the imagery data is not influenced by any other variable, even if acquired on the same area from other sources. For example, given the I_1 and I_2 images, acquired, respectively, by the sources 1 and 2, if we know that a given pixel corresponds to a "flooded bare soil area" in image 2, the knowledge of the I_1 value (and of any other variables) no longer gives us information about I_2. Formally, for the numerator in (3), the following relation holds:

$$\begin{aligned}
&\sum_{C_1}\cdots\sum_{C_M} p(F=\text{flood}, I_1=I_1^*,\ldots,I_M=I_M^*,\\
&\qquad C_1,\ldots,C_M, A_1=A_1^*,\ldots,A_N=A_N^*)\\
&=\sum_{C_1}\cdots\sum_{C_M} p(I_1=I_1^*,\ldots,I_M=I_M^*|F=\text{flood},\\
&\qquad C_1,\ldots,C_M, A_1=A_1^*,\ldots,A_N=A_N^*)\\
&\qquad \cdot p(F=\text{flood}, C_1,\ldots,C_M, A_1=A_1^*,\ldots,A_N=A_N^*)\\
&=\sum_{C_1} p(I_1=I_1^*|C_1)\cdot\ldots\cdot\sum_{C_M} p(I_M=I_M^*|C_M)\\
&\qquad \cdot p(F=\text{flood}, C_1,\ldots,C_M, A_1=A_1^*,\ldots,A_N=A_N^*)
\end{aligned} \tag{4}$$

In the case of remotely sensed images, it is worth noting that InSAR coherence carries information about the temporal stability of the microscopic arrangement of the individual scatterers on the terrain, so it measures the degree of similarity of two SAR images, at the level of the speckle pattern of single pixels. So, strictly, the coherence computed between two SAR acquisitions is not independent from them. However, although InSAR coherence cannot be considered as a channel independent from SAR intensity, their probabilities satisfy a conditional independence property.

- Each C_i term, with $i = 1,\ldots,M$, is conditionally independent of the $A_1,\ldots,A_N$ nodes and of each other $C_j, j \neq i$ term in the network given its parent F. So, if we know that a given pixel corresponds to a "flooded area," the knowledge of the ancillary data and of each other C_j class (i.e., vegetated flooded area) no longer gives us information about the C_i one. Formally:

$$\begin{aligned}
&p(F=\text{flood}, C_1,\ldots,C_M, A_1=A_1^*,\ldots,A_N=A_N^*)\\
&=p(C_1,\ldots,C_M|F=\text{flood},A_1,\ldots,A_N)p(F=\text{flood},A_1,\ldots,A_N)\\
&=p(C_1|F=\text{flood})\cdot\ldots\cdot p(C_M|F=\text{flood})p(F=\text{flood},A_1,\ldots,A_N)
\end{aligned} \tag{5}$$

- Finally, it is important to point out that the N ancillary information sources, available on the analyzed scene, are mathematically not independent among each other.

By considering these assumptions, Eq. (3) can be rewritten in the following form:

$$\begin{aligned}
&p(F=\text{flood}|I_1=I_1^*,\ldots,I_M=I_M^*,A_1=A_1^*,\ldots,A_N=A_N^*)\\
&=\frac{\prod_{i=1}^{M} p\left[\sum_{j=1}^{K_i} p(I_i=I_i^*|C_{i,j})p(C_{i,j}|F=\text{flood})\right]p(F=\text{flood}|A_1=A_1^*,\ldots,A_N=A_N^*)}{\sum_F\left\{\prod_{i=1}^{M} p\left[\sum_{j=1}^{K_i} p(I_i=I_i^*|C_{i,j})p(C_{i,j}|F)\right]p(F|A_1=A_1^*,\ldots,A_N=A_N^*)\right\}}
\end{aligned} \tag{6}$$

2.4 Flood Probability Computation

Each term in Eq. (6) has been operationally computed in the manner described in the following. However, it is worth noting that each term could be computed in a different way, by using different algorithms, models, and ideas.

The $p(I_i = I_i^*|C_{i,j})$ terms. Each $p\left(I_i = I_i^*|C_{i,j}\right)$ is given as an m_i-dimensional probability function generated by a mixture of K_i distributions $j = 1, \ldots, K_i$. Here, K_i Gaussian distributions $\mathcal{N}(\mu_{C_{i,j}}, \Sigma_{C_{i,j}}), j = 1, \ldots, K_i$, are considered. The parameters $\{\mu_{C_{i,j}}, \Sigma_{C_{i,j}}\}$ of each function can be automatically computed by applying a clustering algorithm [3]. In the present case, a K-means algorithm is considered [3], which can be substituted by any other clustering procedure. In particular, it is important to point out that the K-means algorithm works considering, for each pixel in the input images, only the spectral information, while spatial context information is not used. Thus, an interesting option could be to consider a spatial context-based clustering method, especially when very high resolution images are considered in input, i.e., when it is reasonable the assumption that a pixel is surrounded by pixels of the same class [5]. Moreover, while it is common to assume Gaussian statistics for most data sets coming from optical sensors, it is worth noting that full-resolution SAR and InSAR data are usually modeled through different distribution functions, such as gamma or Wishart [19, 31, 34]: however, if the number of looks of the filtered SAR images is sufficiently high, both the SAR intensity and InSAR coherence pdfs can be well approximated with Gaussian functions. Thus, the abovementioned use of Gaussian multivariate pdfs is justified assuming, as is very often the case, that a consistent multilook filtering is applied to the data in the preprocessing steps. The number K_i of each hidden variable C_i (i.e., the number of clusters in the K-means algorithm) can be set by the user if some a priori knowledge about the analyzed area is available; otherwise, it can be determined by a trial-and-error procedure, in order to find the value that provides a good representation of the different classes believed to be present in the scene. This has to be chosen by considering that it is mandatory to avoid under-clustering, but over-clustering is not a problem. In fact, if the number of clusters is too low, there could be mixed clusters, composed of objects with different spectral signatures. In this case, the final result could be severely compromised. On the contrary, if the number of clusters is higher than the best one, there could be several clusters associated with the same spectral signature, but this does not compromise the final result.

The $p(C_{i,j}|F)$ terms. The probabilities $p(C_{i,j}|F), j = 1, \ldots, K_i$, for each independent source data set, $i = 1, \ldots, M$, are computed by applying the Bayes rule:

$$p(C_{i,j}|F) = \frac{p(F|C_{i,j})p(C_{i,j})}{p(F)}, \tag{7}$$

where the $p(C_{i,j})$ terms, derived from the previous $p\left(I_i = I_i^* | C_{i,j}\right)$ computation, are given by the relative numbers of pixels in each of the $C_{i,j}$ clusters and the $p(F|C_{i,j})$ are assigned by the user. These are crucial terms because they disclose the relation between each $C_{i,j}$ object and the flood condition, i.e., they model the electromagnetic response of different classes in flood and no-flood condition. As mentioned, the signal return from some land cover classes can be very complex and strongly depends on acquisition, system, and environmental parameters; moreover, the possible presence of water (with land cover classes partially submerged) makes the modeling even more difficult. Several studies have been conducted about the characterization of the spectral behavior of different semantic classes in different ranges of the e.m. spectrum [6, 15]. A few of these consider the backscatter behavior of some land cover classes in flooded conditions [21, 25]. Here, a simple model is considered, which computes the $p(F = \text{flood}|C_{i,j})$ terms by comparing the $\mu_{C_{i,j}}$ values (previously computed) with empirical thresholds, suitably determined. This model is based on the following considerations about different types of images.

In optical images, the reflectance of clean water is maximum at the blue end of the e.m. spectrum and decreases as wavelength increases. Turbid water typically contains suspended sediment, which increases the reflectance in the red end of the spectrum, accounting for its brownish appearance. The reflectance of bare soil depends on its composition, but generally it increases monotonically with increasing wavelength. Vegetation can be identified by the high NIR but generally low visible reflectance. As mentioned in Chap. 1, several indices have been proposed to highlight the specific presence of water in optical images. Nevertheless, red and near-infrared are among the most diffused bands in use with optical sensors. For these reasons, a model considering as input the Normalized Difference Vegetation Index (NDVI) [15] is considered here. NDVI is defined as:

$$\text{NDVI} = \frac{\rho_{\text{NIR}} - \rho_{\text{R}}}{\rho_{\text{NIR}} + \rho_{\text{R}}}, \tag{8}$$

where ρ_{NIR} and ρ_{R} are the reflectances in the NIR and red band, respectively. Often, the use of the radiances instead of the reflectances in NDVI computation is justified by assuming constant atmospheric and other effects on the two close spectral bands. A suitable threshold on the NDVI value allows to distinguish between water from other land cover classes in conditions of low-medium turbidity, while often very turbid water and mud cannot be safely distinguished from bare soils [15]. If more images, acquired before and during the flood, are available, it is possible in principle to distinguish permanent water from flooded areas.

When InSAR coherence is considered, low values are commonly associated with water. So, by setting a threshold value, it is possible to separate water areas from other land cover classes. However, an overestimation of flooded area is possible because other classes (such as forests or other vegetated areas) may also exhibit typically low coherence value. If more coherence images (acquired before, during, and/or after the event) are available, it is possible in principle to distinguish areas showing low coherence values even in dry conditions (typically vegetated terrains), from flooded areas, which usually exhibit a reduction of coherence during the inundation. It should not be forgotten that the InSAR coherence information is a change measurement and thus refers to the changes occurred on the surface between the two acquisitions defining the InSAR image pair. Here, we assign coherence information to both the dates of acquisition of the images composing the InSAR pair.

In SAR backscattering intensity images, smooth open water can be easily detected when the flat water surface acts as a specular reflector, and so the backscattered signal is very low. A threshold value can then be used to separate water from other classes. Moreover, by considering more images (acquired, e.g., before and during or after the inundation), it is possible to separate permanent water from flooded areas. As we have seen, other typical behaviors can be isolated if a time series of SAR images is available. For example, over partially flooded crop land, in many cases the backscattered signal increases during an inundation, due to double-bounce effects [30]; however, these effects are difficult to be modeled because they show a great variability depending on crop characteristics (plant structure, plant and leaf density, relative height with respect to water level, growth stage, etc.) and also on the acquisition system (such as frequency band, acquisition angle, polarization). A similar situation may occur in urban areas, where other effects (such as layover) could heavily influence the flood detection. More details can be found in [25]. Therefore, in order to consider the double-bounce effects, it is important to have some experience with backscattering images and to have some information on the land cover.

The $p(F|A_1,\ldots,A_N)$ term. The conditional probability $p(F|A_1,\ldots,A_N)$, representing the a priori probability of flood based only on the ancillary data, can be modeled in different ways. We recall that the N ancillary information sources, available on the analyzed scene, can be mathematically not independent among each other. So, a function depending on the whole set $A_k, k = 1,\ldots,N$ has to be considered. This corresponds to consider a single variable that is a combination of the A_k. However, it is possible to consider each $p(F|A_k), k = 1,\ldots,N$ probability term as independent from the other ones, assuming that the F variable is not directly observed [18]. This position is a useful simplification. In the following, both cases have been considered.

In any case this probability term can be modeled in various ways. Here, two different monotonic functions bounded in the [0,1] interval are considered:

1. A sigmoid function:

$$p(F|A) = \frac{1}{1+\exp[-\sigma\cdot(A-\mu)]},$$

where μ is the A value corresponding to $p(F|A) = 0.5$ and σ is the steepness of the curve. These parameters are set by the user.

2. A piecewise linear function:

$$p(F|A) = \begin{cases} 0 & \text{if} A \leq g_1; \\ \dfrac{A - g_1}{g_2 - g_1} & \text{if} g_1 < A < g_2; \\ 1 & \text{if} A \geq g_2, \end{cases}$$

where the g_1 and g_2 are user parameters.

3 Experimental Results

3.1 *Study Area and Test Events*

Several flood events occurred in the last years in the Basilicata region (Southern Italy) and, in particular, in its coastal area on the Ionian Sea, due to its intrinsic lithological and geomorphological properties [4]. In this area, several rivers are flowing (see Fig. 3). The surrounding land cover is mainly agricultural, characterized by cereal and vegetable crops or fruit shrubs, while an extended area near the river mouths and along the coast is occupied by pine forest stands [11]. The frequency of extreme events has increased, in the last 5 years [24], causing several flooding episodes, even two–three in the same year, with great damage for agriculture and industries. This process is enhanced by the agricultural land use of the alluvial plains, which has reduced the extent of the natural river beds, increasing flood risk. We focus on an inundation event occurred on Nov. 2–6, 2010. The event was due to a rainfall occurred in the late morning of November 2. The rainfall lasted about 6 h, with most of it concentrated in 2 h (10 a.m.–12 a.m.). The Bradano River provoked inundations that affected several areas during the night between November 2 and 3, 2010. The peak flow at the gauging station SS106 (close to the outlet, see Fig. 4) was recorded on November 3, at 6 a.m. [29]. This peak produced an inundation that propagated in the surrounding areas, producing floods during Nov. 3–6.

3.2 *Available Data Sets*

The acquisition dates of the available input imagery data are reported in Table 1. For the considered event, there are images (reported in bold characters in Table 1) acquired during the inundation, thus offering a useful observation data set to follow the temporal evolution of the flood wave phenomenon, while the other scenes may act as reference data.

- **SAR intensity data.** All the SAR images are acquired in stripmap mode by the COSMO-SkyMed Italian satellite constellation, with a ground pixel size of approximately 3×3 m^2, in the same geometry (ascending), polarization

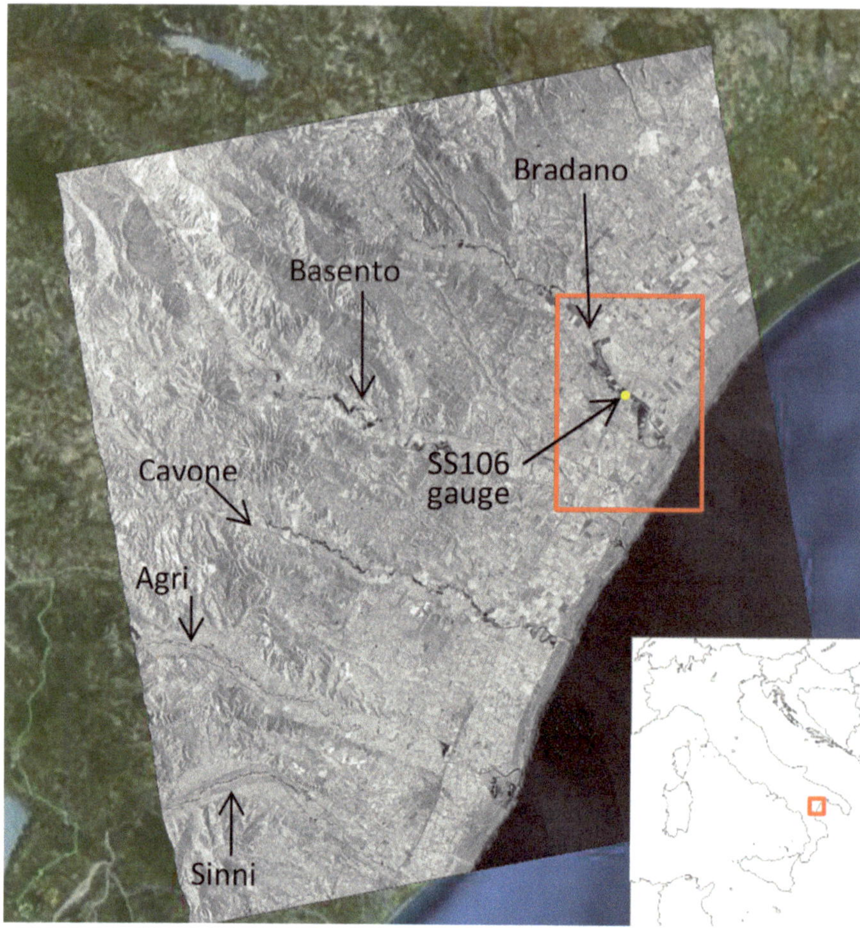

Fig. 3 Location of the test site. *Bottom-right inset* shows large-scale location of the Basilicata coastal area (*small red square*). River names, approximate position of the SS106 streamflow station, and test area (*red rectangle*) are indicated on the georeferenced map of one of the SAR images used in the processing. Background is an optical image from Google Earth™

Table 1 Acquisition dates of available imagery data. Bold text indicates co-event acquisitions

Event date	SAR images	InSAR pairs	Optical images
2010	October 2	October 2–3	**November 6**
	October 3	**November 3–4**	
	October 10		
	November 3		
	November 4		
	November 11		

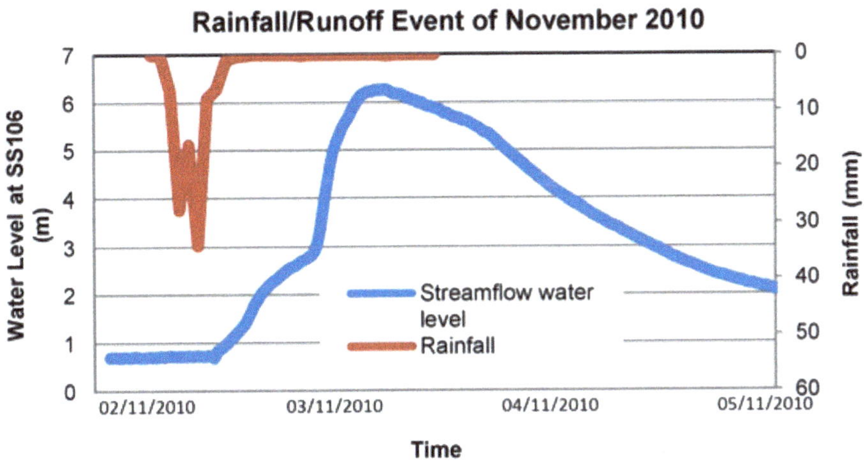

Fig. 4 Description of the flood event recorded during November 2–4, 2010. The graph describes the temporal evolution of the water level recorded at the SS106 streamflow station and of the rainfall amount averaged over the basin area

(VV), and incidence angle (38°). Preprocessing included standard SAR intensity calibration. All the images were precisely co-registered at sub-pixel precision on a common master, using height information from SRTM data [22]. A filtering step was performed using a nonlocal approach [10]. Final filtered intensity images were geocoded to a common reference frame, thus obtaining regular-grid UTM maps with a posting of 3 m in both dimensions.

- **InSAR coherence data.** The coherence images have been obtained from the 1-day interferometric SAR pairs among the previously described data. Spatial baselines for the data set are contained within a few hundred meters interval, which is an order of magnitude smaller than the critical baseline value for COSMO-SkyMed data. This ensures that geometric decorrelation [36] is negligible in the considered data set, so that 1-day separation InSAR pairs exhibit coherence information which is mainly affected by temporal variations between the two acquisitions. Interferometric processing was thus performed through the DORIS open source software [17]. Coherence was computed on a 11×11 pixels sliding window, previously removing the systematic topographic phase from the data. These images were also geocoded to the same reference frame as the intensity ones.
- **Optical data.** A RapidEye [28] optical image, at a ground resolution of 5 m, acquired on November 6, 2010, which presents sufficient cloud-free information around the study area, was available for the flood event considered. It is acquired in five spectral bands, of which the ones corresponding to the red and near-infrared intervals were used to obtain the NDVI according to Eq. (8), with radiance values in place of reflectance. The NDVI maps were geocoded to the

same reference frame of the other images, covering part of the test area covered by the SAR data.

- **Ancillary data**. The following ancillary data sources were considered:
  - **Distance from the river.** The Euclidean distance from the Bradano River bed was used, based on the heuristic principle that the higher the distance by the river, the lower the probability that a pixel could belong to the flooded class.
  - **Geomorphic flooding index (GFI).** This ancillary datum is computed from the definition in [20, 32]:

$$\mathrm{GFI} = \ln(h_r/H), \tag{9}$$

where h_r is a function of the contributing area[1] A_r in the nearest section of the drainage network hydrologically connected to the point under exam, and H is the elevation difference between the cell under exam and the final point of the above-identified path. In this index, the parameter h_r is representative of the probable water level in a cross section of the river hydraulically connected to the point, and it is computed as a power law of A_r with exponent set equal to 0.3. The variable GFI is introduced to assign correct a priori flood probability to zones taking into account the structure of the river network, the relative magnitude of each stream link, and also the relative elevation of the surrounding area.

3.3 *Flood Map Production: SAR Intensity Data and Distance from the River*

First of all, the flood event has been analyzed by considering only the six multi-temporal SAR intensity images listed in Table 1 and the distance from the Bradano River as ancillary datum. The corresponding BN scheme is reported in Fig. 5.

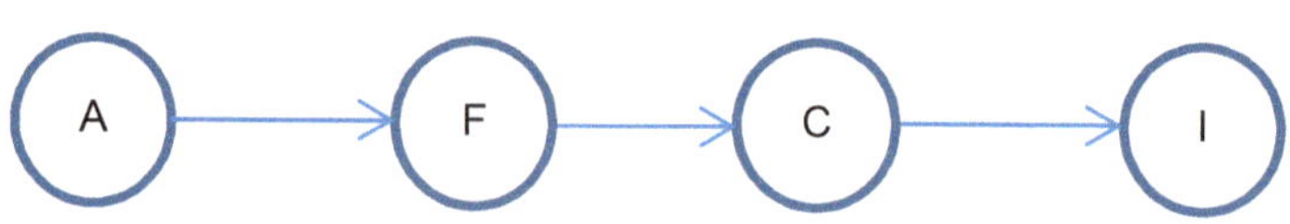

Fig. 5 BN scheme in which only the SAR intensity data and the distance from the river have been considered

[1]The contributing area, or drainage area, of a given river point is the sum of the hydrologically connected area upslope that point. A contributing area can be generally calculated for every landscape point by iteratively summing unitary "parcels" or pixels upslope a given point (see, e.g., [2]).

Table 2 Probability values $P(F|C)$ assigned to clusters for the Nov. 3 and 4, 2010, test cases. Centroid arrays consist of the six backscattering intensities (in dB) in all the available acquisitions, in temporal order from left to right (Oct. 2, 3, and 10, Nov. 3, 4, and 11)

Centroid value (dB)						$P(F\|C)$	
I_1	I_2	I_3	I_4	I_5	I_6	03/11/10	04/11/10
−9.17	−9.04	−9.62	−16.74	−7.32	−8.68	0.90	0.10
−8.03	−8.89	−9.47	−9.45	−16.59	−9.55	0.10	0.90
−9.02	−9.34	−10.05	−13.93	−17.48	−17.32	0.90	0.90
−7.00	−7.40	−7.79	−5.25	−4.79	−6.59	0.70	0.70
−6.97	−8.13	−8.79	−6.56	−6.24	−8.59	0.70	0.70
−17.11	−17.36	−17.56	−16.76	−16.68	−17.08	0.10	0.10
−1.65	−2.16	−2.48	−0.04	0.19	−1.96	0.05	0.05

In particular, the conditional probability $p(I|C)$ has been computed as a multivariate, six-dimensional probability function, composed of a mixture of Gaussian distributions. As mentioned above, the objective of multi-temporal imagery segmentation is to take into account the behavior of different land cover types in presence of inundation and with different levels of water height. We have found that $K_1 = 32$ is a number of clusters sufficient to identify univocally all the different types of multi-temporal scattering classes present in the data. Of these, a subset represents objects of interest for the specific flood monitoring experiment.

In order to set the corresponding $P(F|C)$ values, the temporal trends of the mean value of SAR intensity (σ^0) in each cluster have been considered. In Table 2, some $P(F|C)$ values, related to the flood dates of November 3 and 4, are reported as an example.

In particular, high probability values are assigned to clusters that exhibit significant variations of their mean values for the images acquired during the flood event. For example, the trend in the first row of Table 2 is characterized by very small σ^0 values on November 3, while, before and after this date, the radar return is higher: regions showing this multi-temporal signature were assumed to be flooded on November 3 only. As a consequence, $P(F|C) = 0.9$ for November 3 and $P(F|C) = 0.1$ for the next day. Similarly, the trend reported in the second row shows very low σ^0 values only on November 4: we assume it characterizes regions flooded on November 4 only. The trend in the third row exhibits low σ^0 values on the three dates of November 3, 4, and 11: it can be hypothesized that such signatures represent areas flooded throughout all such periods. The following two trends are characterized by values around −6 dB on the first dates, increasing on November 3 and 4: this can be explained by considering the double-bounce effect [25]. Actually, there are other four clusters, exhibiting the abovementioned σ^0 value trend and very likely reflecting the complexity of the phenomena and the variations in radar response due, e.g., to plant height, water depth, and their relative values. To these clusters, $P(F|C) = 0.7$ is assigned, considering that other "objects" (such as wet bare soils) on the ground may show the same behavior. The trends in the last

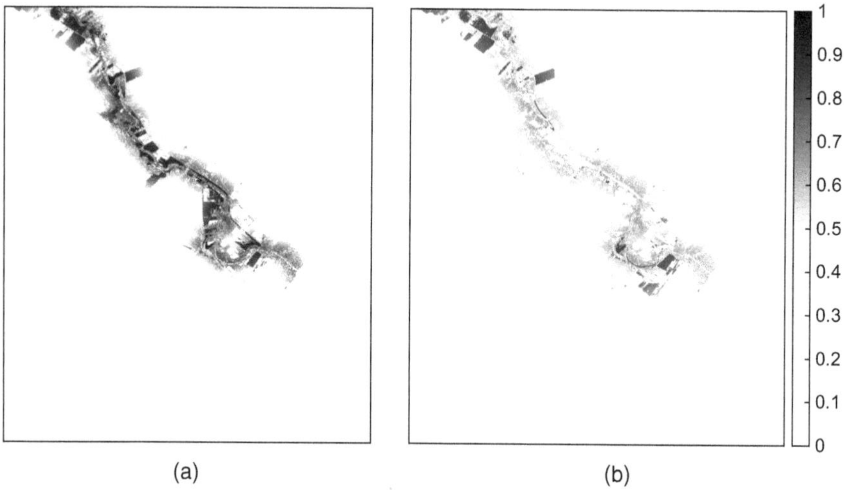

Fig. 6 Probabilistic flood maps produced by the BN approach for the dates of Nov. 3 (**a**) and Nov. 4, 2010 (**b**), on the area highlighted by the *red rectangle* in Fig. 3

two rows of Table 2 are associated to no-flood conditions, and consequently low probability values were assigned. Generally, very low probabilities were assigned to all the clusters showing constantly high or constantly low intensity values.

The probability $p(F|A)$, with A given by the distance d from the river in this case, has been computed as a sigmoid function of d, which decays with different steepness in each of the monitored days, according to some information available on the considered event; for this reason, the parameters reported in the second column of Table 5 have been used. The flood maps depicted in Fig. 6 represent the final $p(F|I, A) > 0.5$ posterior probability of flood, computed on the basis of the available evidence.

To have an idea of the performances of the method, the two BN-derived flood maps of the 2010 event have been compared with other reference flood maps, independently obtained from hydraulic modeling (see [29] for details). An overall accuracy $A = 92\%$ and $A = 96\%$ is obtained, respectively, for the Nov. 3 map and for the Nov. 4 map, considering the binary maps with $p > 0.5$. Actually, the sets of flood and no-flood pixels are very unbalanced on the maps (the extension of the flooded area is relatively small with respect to the total analyzed area), so the true positive rate (TPR) and false positive rate (FPR) values are also considered to give more objective evaluations. They are defined as:

$$\mathrm{TPR} = \frac{\mathrm{TP}}{\mathrm{TP} + \mathrm{FN}}; \tag{10}$$

$$\mathrm{FPR} = \frac{\mathrm{FP}}{\mathrm{FP} + \mathrm{TN}}; \tag{11}$$

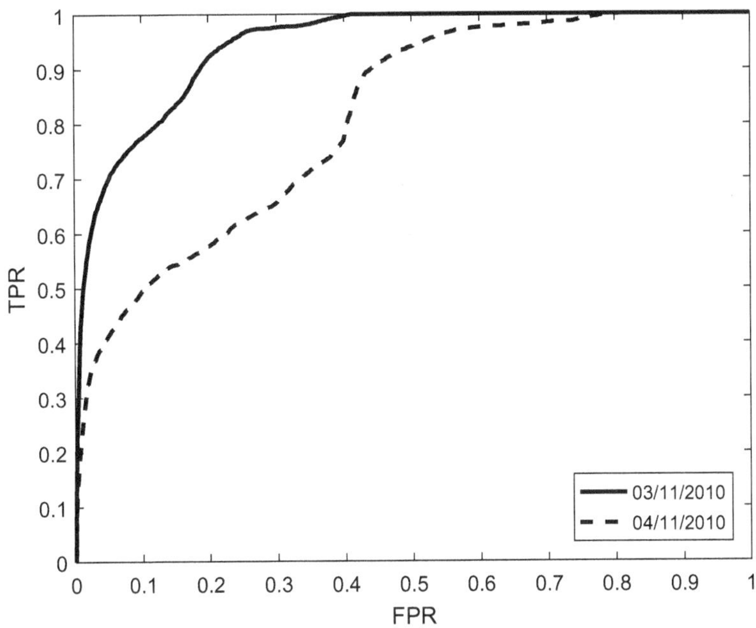

Fig. 7 ROC curves obtained by comparing the flood probability maps in Fig. 6 with the reference data for Nov. 3 and Nov. 4, respectively

where TP is the number of true positives, i.e., the actual flood data that are correctly classified; FP is the number of false positives, i.e., negative data classified as positive; TN is the number of true negatives, i.e., the actual no-flood data that are correctly classified; and FN is the number of false negatives, i.e., positive data classified as negative. In Fig. 7, the corresponding receiver operator characteristic (ROC) curves, obtained by using increasing threshold P values for the BN-derived probability maps, are reported: a good agreement with the reference maps has been obtained, with areas under the curve (AUC) values for the Nov. 3 map and for the Nov. 4 map equal to 0.91 and 0.81, respectively. As can be seen, these figures correspond to moderately prudent estimators, with reasonable sensitivities sided by very good specificities.

Although only some of the available images have been used, promising results have been obtained. However, there are some critical behaviors that require a more detailed analysis. In particular, the colored polygons in Fig. 8 have been considered. The red polygon is a false alarm example: it shows a mean probability value, computed on its area, equal to $\overline{P}(F = \text{flood}|I, A) = 0.73$, although it is modeled as not flooded by the abovementioned hydrogeological model. In fact, its SAR intensity mean values, computed on the polygon area, for the dates corresponding to the flood event, are lower than the values measured in the other available images, as reported in Table 3. Although the SAR intensity decrease, exhibited in this area during the event, is not comparable with the ones observed in actual flooded areas, it was labeled by the BN network with high posterior probability values.

The other three polygons highlighted in Fig. 8 correspond to areas covered by vineyards, as it can be observed by two optical images, freely available on Google Earth, acquired before (April 4, 2010) and after the event (March 12, 2011) (see [7] for further details). They are very likely examples of missed identifications; in fact, the abovementioned hydrogeological model predicts water levels in these areas in the range 50–115 cm. In these cases, by considering the typical vineyard structure, it is reasonable that a double-bounce effect occurred in these areas, although the SAR backscatter increase is lower than the ones exhibited in other areas characterized by the same effect.

3.4 Flood Map Production: SAR Intensity and InSAR Coherence Data and Distance from the River

In order to overcome the limitations previously reported, a further experiment has been performed, where the SAR backscatter and InSAR coherence data have been considered, and treated as conditionally independent sources of information. The BN scheme for this new experiment is reported in Fig. 9.

In particular, the I_1 variable corresponds to the six-dimensional vector previously considered, while the I_2 variable is the two-dimensional vector obtained from the two InSAR coherence images. Also, the only ancillary data source considered is again the distance map from the Bradano river. In this case, the BN structure shows its great advantage given by its modularity. In fact, thanks to the independence relations between new and old variables, the probability values assigned to the clusters related to the other nodes can be reused, and only the conditional probability terms due to the new variable have to be computed. In particular, the $P(I_1|F)$ and the $P(F|A)$ probability values previously computed have been used, with the same $P(F|C_{1,j})$ probabilities (see Table 2) and the same parameters (see the second column in Table 5), respectively.

The segmentation of the multi-temporal InSAR coherence data set was performed in this case, as mentioned, independently from the intensity data. On most types of natural terrain, coherence is expected to be low on areas which have been inundated for some time during the interval between the two InSAR acquisitions, but this signature is also a characteristic of other land cover classes such as permanent water and forests, also present in the analyzed scene. The availability of two InSAR coherence images is particularly useful in the interpretation of flooded and not flooded areas, because it allows to separate areas exhibiting low coherence values even in normal conditions (permanent water or forests) and areas subjected to a decrease of coherence during the inundation period (flooded areas). On the other hand, it should be noted that the coherence value on flooded areas is expected to be low even if they have been covered by water only on one of the two images

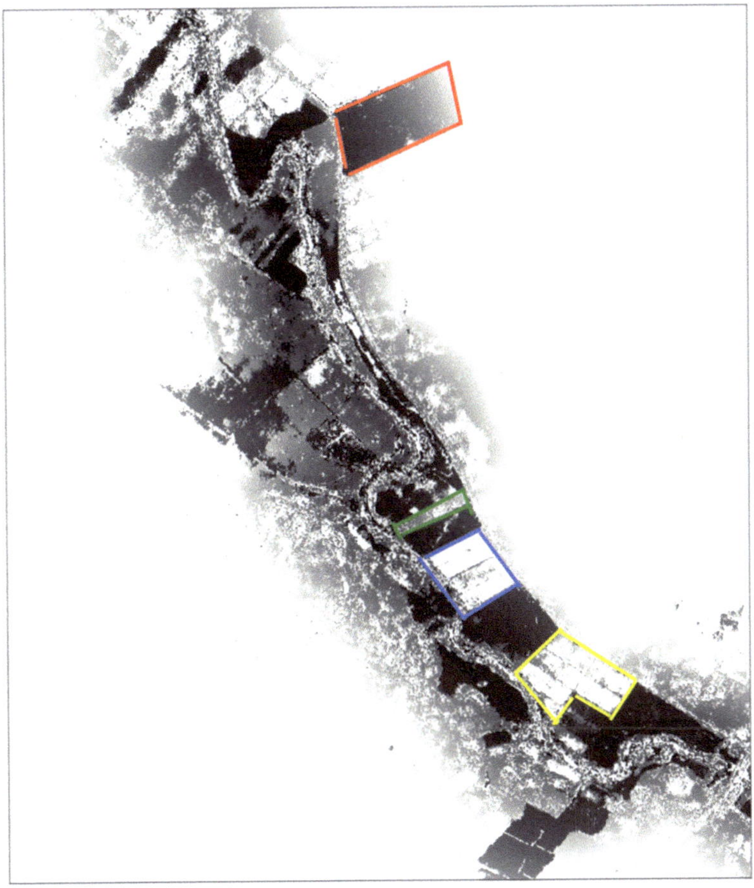

Fig. 8 A detail of the probabilistic flood map for Nov. 3, shown in Fig. 6a, with some areas highlighted by using different colors

Table 3 Mean probability values $\overline{P}(F|I, A)$ assigned to each polygon in Fig. 8 for the Nov. 3, 2010 test case. Centroid arrays consist of the six mean backscattering intensities (in dB), computed on each polygon, in the whole available acquisitions, in temporal order from left to right (Oct. 2, 3, 10, Nov. 3, 4, and 11)

	Centroid value (dB)						
Polygon	I_1	I_2	I_3	I_4	I_5	I_6	$\overline{P}(F\|A, I)$
Red	−7.15	−8.88	−10.42	−13.67	−13.06	−14.79	0.73
Green	−9.60	−9.24	−9.35	−9.98	−8.39	−7.41	0.53
Blue	−6.81	−6.15	−7.52	−6.72	−5.98	−7.56	0.26
Yellow	−7.25	−6.53	−7.34	−7.66	−6.52	−7.36	0.31

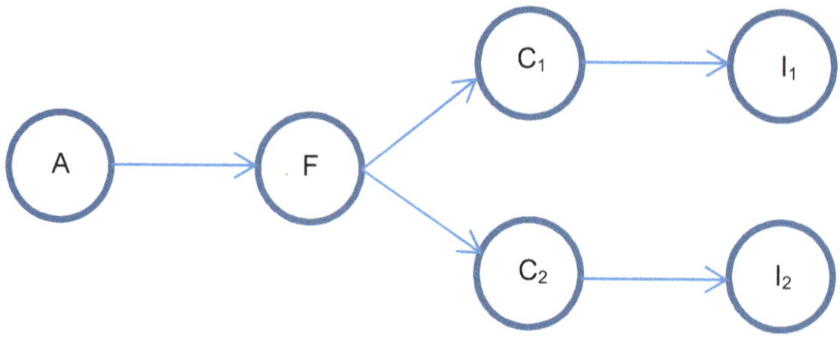

Fig. 9 BN scheme in which the SAR backscatter (I_1) and InSAR coherence (I_2) data are used, considered as conditionally independent

Table 4 Probability values $P(F|C_2)$ assigned to clusters for the Nov. 3, 2010, test case. Centroid arrays consist of the two coherence values of the pairs Oct 2–3 and Nov. 3–4, respectively

Centroid value		$P(F\|C_2)$
I_1	I_2	
0.15	0.13	0.5
0.59	0.20	0.9
0.68	0.69	0.1

forming the InSAR pair, or even for any time interval between the first and the second sensor passage, but not at the time of the two acquisitions. Since the InSAR coherence image is inevitably associated with both the acquisition dates of the pair, in the above cases, this may lead to an overestimation of the time interval during which a particular area was interested by floodwater. This aspect could be a drawback in the following flood dynamics, but it can be overcome when the coherence information is suitably merged with other kinds of data. Again, various tests were performed in this case, finding that $K_2 = 8$ clusters give a reliable representation of the analyzed scene, showing several distinguishable behaviors. In particular, three different temporal trends can be pointed out: clusters characterized by constantly high coherence value, clusters having constantly low coherence value, and clusters showing a significant reduction of coherence during the flood event.

The probabilities $p(C_2|F)$ have been computed by applying the Bayes rule, with the $P(F|C_2)$ terms assigned by the user. An example of some assigned $P(F|C_2)$ is reported in Table 4. In particular, the clusters showing significant decrease of coherence are assigned high values of $P(F|C_2)$; clusters exhibiting constantly low coherence values are given lower values. Finally, all the other clusters, showing constantly high intensity value and high coherence value, are given very low probabilities. It is important to note that, for example, a given pixel can be characterized by an intermediate value of $P(F|C_1)$ and a high value of $P(F|C_2)$, because the two kinds of data have been considered as independent.

A probability value to be flooded is assigned to each pixel in the analyzed scene by using Eq. (3), i.e., by fusing SAR intensity data, InSAR coherence data, and distance from the river. The flood maps depicted in Fig. 10 represent the final

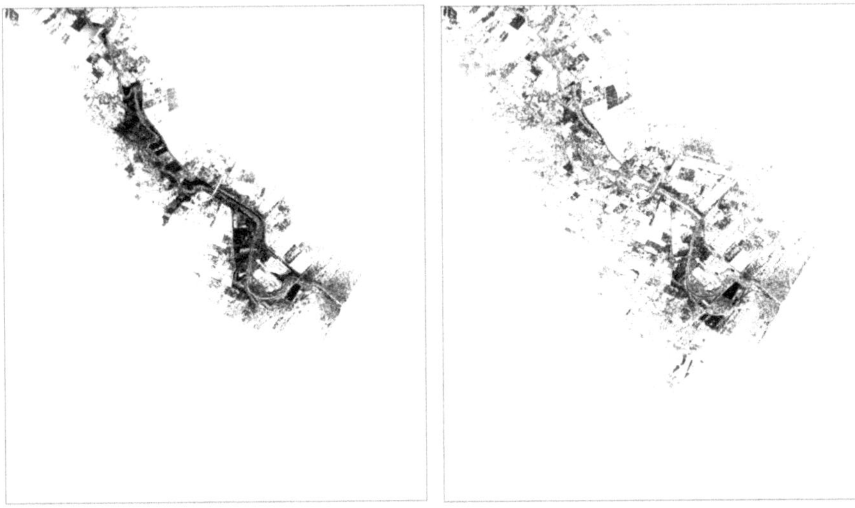

(a) (b)

Fig. 10 Probabilistic flood maps produced by the BN approach through the scheme in Fig. 9, for the dates of Nov. 3 (**a**) and Nov. 4, 2010 (**b**). Location and grayscale are the same as in Fig. 6

posterior probabilities of flood. The two probabilistic flood maps have been again compared with the same reference flood maps previously cited, independently obtained from hydraulic model [29]. The best overall accuracy values, obtained by varying the probability threshold values, are respectively, $A = 92\%$ and $A = 96\%$. In Fig. 11, the corresponding ROC curves are reported: a good agreement with the reference maps has been obtained, with AUC values for the November 3 and 4 maps, respectively, equal to 0.95 and 0.83. Moreover, all the four polygons previously considered are correctly classified.

3.5 Flood Maps Production: SAR Intensity, InSAR Coherence, Optical Data, and Distance from the River

We have then also considered the information extracted from an optical image, i.e., a RapidEye multispectral image [8, 28], whose red and infrared bands were used to compute, for each pixel, the corresponding NDVI index. We used this image together with the other data used in the preceding experimental setup, with the additional optical image node considered, as in Fig. 9, conditionally independent from the other SAR and InSAR image nodes. In this case, the optical image has been used only to follow the flood dynamics, because it has been acquired in a different date with respect to the SAR acquisition. In principle, optical or other data, acquired in dry conditions, could also be used as additional ancillary ("A") nodes within the

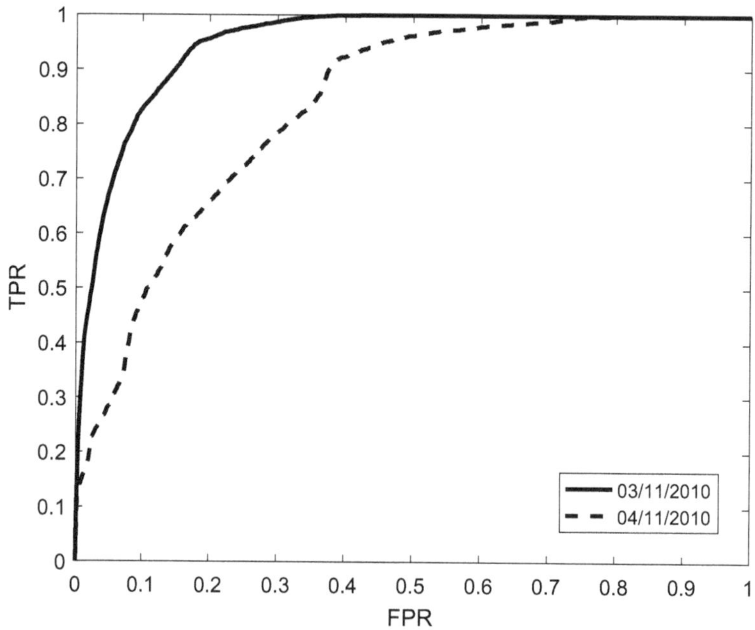

Fig. 11 ROC curves obtained by comparing the flood probability maps in Fig. 10 with the reference data for Nov. 3 and Nov. 4, respectively

BN scheme in Fig. 9 or indirectly to extract information concerning the land cover of the analyzed area and thus, for example, to reassign the probabilities $P(F|C_1)$ in order to identify, e.g., vegetated fields potentially causing double-bounce.

As mentioned, thanks to the BN modularity, only the conditional probabilities $P(I_3|C_3)$, with I_3 = NDVI, have been here computed, as mono-dimensional probability functions generated by a mixture of $K_3 = 16$ Gaussian distributions, with parameters obtained through K-means clustering, as specified above. The probabilities $P(F|C_3)$ have been assigned by considering the centroid value μ_{C_3} of each intermediate class: areas with negative NDVI values have been considered as most likely flooded. The probability $P(F|A)$ has been computed again as a sigmoid function of the distance $A = d$ from the river, which decays with different steepness in each of the monitored days, as reported in the second column of Table 5.

The flood map for November 6 is depicted in Fig. 12, by considering a threshold of 0.5 on the final probability map. From Figs. 12 and 10, the temporal evolution of the phenomenon can be inferred, with fields with high probability of being flooded exhibiting maximum values in wide areas close to the river course on November 3, then on areas farther away from the river on November 4, and finally a few ponds of stagnating water appearing as black areas on the map for November 6.

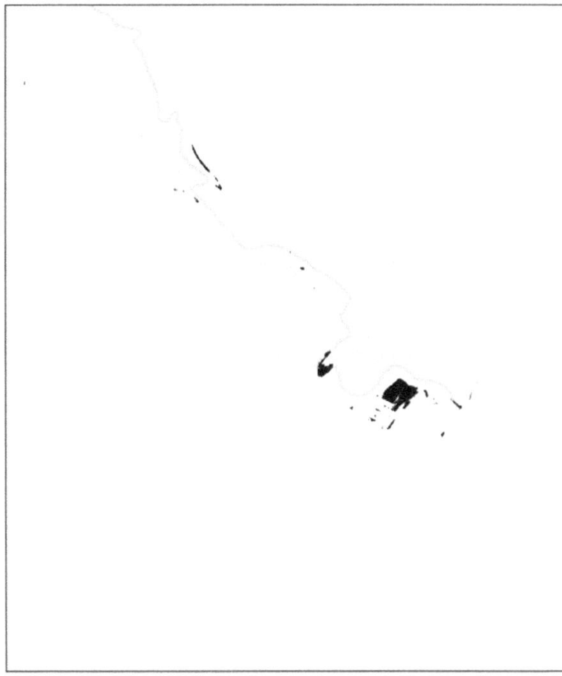

Fig. 12 Flood maps produced by BN for Nov. 6, 2010, with the BN scheme described in Sect. 3.5. Location and grayscale are the same as in Fig. 6

3.6 Flood Maps Production: SAR Intensity, InSAR Coherence, Optical Data, Distance from the River, and Geomorphic Ancillary Data

Finally, the same imagery data have been considered as the preceding experiment, but another ancillary information source, i.e., the GFI, has been added. Two different relations between the two ancillary data sources have been explored. First of all, the conditional probability function $P(F|A_1, A_2)$ has been modeled as a sigmoid. In our case, $P(F|A_1, A_2)$ is a function of two independent variables, for which the following linear relation is considered:

$$A = t_1 \cdot A_1 + t_2 \cdot A_2 \tag{12}$$

where t_1 and t_2 have been introduced to properly take into account the contribution of each variable. In this way, the two ancillary variables, i.e., the GFI and the distance from the river course, have been combined into a single variable which can be interpreted as a "modified" Euclidean distance, modulated by the weight factors t_1 and t_2 which also serve as a normalization constant to bring both variables within a common interval. These weights, together with the sigmoid function parameters, σ and μ, have to be set by considering information about the considered event. Anyway, this particular relation between variables is not mandatory for the success

Table 5 Function type and parameters for each considered ancillary datum

Ancillary data type	A_1 (Distance d, m)	A_2 (GFI)
Function type	Sigmoid	Piecewise Linear
Parameters for date 2010-11-03	$\sigma = 1/300, \mu = 300$	$g_1 = 1, g_2 = 4$
Parameters for date 2010-11-04	$\sigma = 1/700, \mu = 100$	$g_1 = 3, g_2 = 4$
Parameters for date 2010-11-06	$\sigma = 1/100, \mu = 100$	$g_1 = 3, g_2 = 4$

(a) (b) (c)

Fig. 13 Probability flood maps: (**a**) Nov. 3, (**b**) Nov. 4, (**c**) Nov. 6, 2010, through the BN scheme described in Sect. 3.6. Location and grayscale are the same as in Fig. 6

of the proposed methodology, and other forms could be used. For instance, in order to avoid having to set these parameters, which often require a detailed knowledge of the flood evolution, which is what we want to predict, the independence between ancillary data can be considered. This additional case was then also considered, and the two ancillary information sources were thus considered as conditionally independent. The probability terms were modeled according to a sigmoid function for the variable $A_1 = d$ and according to a piecewise linear function for the variable $A_2 = \text{GFI}$, respectively. The type of function selected and the parameters are reported in Table 5. The flood maps depicted in Fig. 13 represent the final posterior probabilities of flood. The two maps for November 3 and 4 have been compared again with the same reference flood maps, independently obtained by using a hydrologic flood model [20]. No model result was available for the date of November 6. In Fig. 14, the two corresponding receiver operator characteristic (ROC) curves are reported: a fair agreement with the reference maps has been obtained, with areas under the curve (AUC) values for the November 3 and 4 maps equal to 0.94 and 0.86, respectively. The best overall accuracies, obtained by varying the probability threshold values, are, respectively, equal to 93% and 96%, for Nov. 3 and 4.

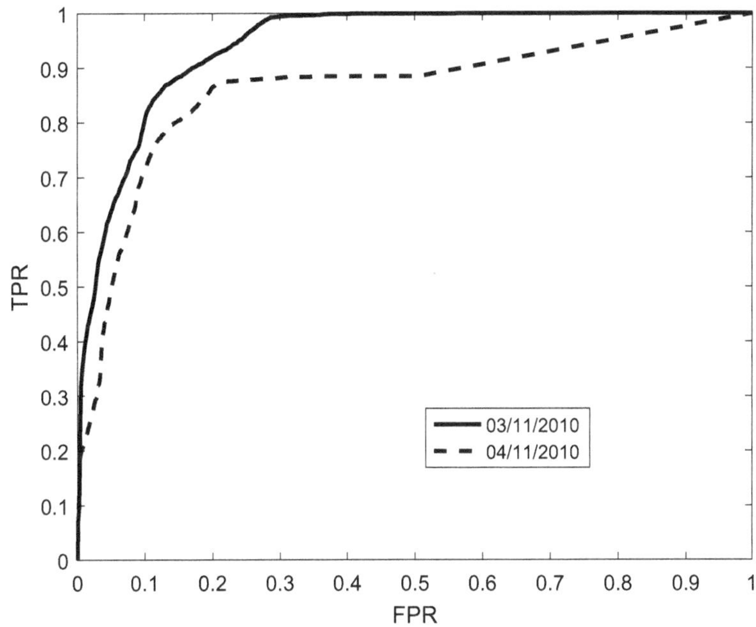

Fig. 14 ROC curves obtained by comparing the probability maps in Fig. 13a,b with the reference maps for Nov. 3 and Nov. 4

4 Conclusions

We have shown an application of a methodology based on Bayesian networks (BN), to perform a data fusion approach of remotely sensed imagery and ancillary data to detect flooded areas. The method exploits the modularity and the simplicity of graphical networks to help set up a model of the independence relations between different variables representing various data sources, such as radar intensity and interferometric coherence, optical multispectral data, and other ancillary data regarding land cover or other geomorphic or hydrological variables. Such model simplifies the assignment of a priori probability values to each variable, in order to derive a series of probabilistic flood maps, one for each data set, thus potentially following temporally a given flood event. Each output map expresses the likelihood of each pixel to be flooded in that acquisition date, on the basis of the various a priori probabilities. Results show consistent capabilities of identification of parameters such as the total area interested by the flood phenomenon at different dates, thus partially overcoming the obstacle constituted by the presence of several scattering/coherence classes corresponding to either different land cover types, which respond differently to the presence of water, or to different temporal phases in the evolution of the phenomenon. Comparisons with benchmark maps obtained by independent flood area delineation methods (hydrological models in this case) show good agreement.

Obviously, more elaborated information, for example, explicit e.m. simulation tools to rigorously infer the expected backscattering response of flooded/nonflooded areas to the incoming e.m. radiation, in both optical and microwave spectral bands, can also be considered to obtain better results. Moreover, different BN schemes (i.e., different ways to merge imagery data) can be considered, in order to exploit the internal dependencies among these variables. In conclusion, BNs appear to be a powerful tool to perform data fusion in the analysis of complex real-world systems, such as natural hazard detection. The goal appears thus to be feasible, of setting up an automated procedure, based on BN, to extract unambiguously flooded areas from satellite images, resolving ambiguities through inference of a priori information from data acquired before or after the event, in absence of flood conditions.

Acknowledgements COSMO-SkyMed images are courtesy of Italian Space Agency. Ing. L. Candela, of the Italian Space Agency (ASI), is kindly acknowledged for support in data acquisition. InSAR processing was performed by Dr. D. O. Nitti of GAP s.r.l.

References

1. Barber, D.: Bayesian Reasoning and Machine Learning. Cambridge University Press, Cambridge (2012). doi:10.1017/CBO9780511804779
2. Beven, K.J., Kirkby, M.J.: A physically based, variable contributing area model of basin hydrology/Un modèle à base physique de zone d'appel variable de l'hydrologie du bassin versant. Hydrol. Sci. Bull. **24**(1), 43–69 (1979). doi:10.1080/02626667909491834
3. Bishop, C.M.C.: Neural Networks for Pattern Recognition. Clarendon Press/Oxford University Press, Oxford/New York (1995)
4. Boenzi, F., Caldara, M., Capolongo, D., Dellino, P., Piccarreta, M., Simone, O.: Late Pleistocene-Holocene landscape evolution in Fossa Bradanica, Basilicata (Southern Italy). Geomorphology **102**(3–4), 297–306 (2008). doi:10.1016/j.geomorph.2008.03.013
5. Bovolo, F., Bruzzone, L.: The time variable in data fusion: a change detection perspective. IEEE Geosci. Remote Sens. Mag. **3**(3), 8–26 (2015). doi:10.1109/MGRS.2015.2443494
6. Bracaglia, M., Ferrazzoli, P., Guerriero, L.: A fully polarimetric multiple scattering model for crops. Remote Sens. Environ. **54**(3), 170–179 (1995). doi:10.1016/0034–4257(95)00151–4
7. D'Addabbo, A., Refice, A., Pasquariello, G., Bovenga, F., Chiaradia, M.T., Nitti, D.O.: A Bayesian network for flood detection. In: 2014 IEEE Geoscience and Remote Sensing Symposium, pp. 3594–3597. IEEE (2014). doi:10.1109/IGARSS.2014.6947260
8. D'Addabbo, A., Refice, A., Pasquariello, G., Lovergine, F.P., Capolongo, D., Manfreda, S.: A Bayesian network for flood detection combining SAR imagery and ancillary data. IEEE Trans. Geosci. Remote Sens. **54**(6), 3612–3625 (2016). doi:10.1109/TGRS.2016.2520487
9. Dasarathy, B.: Information Fusion: what, where, why, when, and how? Inf. Fusion **2**(2), 75–76 (2001). doi:10.1016/S1566–2535(01)00032-X
10. Deledalle, C.A., Duval, V., Salmon, J.: Non-local methods with shape-adaptive patches (NLM-SAP). J. Math. Imaging Vision **43**(2), 103–120 (2012). doi:10.1007/s10851–011-0294-y
11. Fiorentino, M., Manfreda, S., Iacobellis, V.: Peak runoff contributing area as hydrological signature of the probability distribution of floods. Adv. Water Resour. **30**(10), 2123–2134 (2007). doi:10.1016/j.advwatres.2006.11.017
12. Frey, D., Butenuth, M., Straub, D.: Probabilistic graphical models for flood state detection of roads combining imagery and DEM. IEEE Geosci. Remote Sens. Lett. **9**(6), 1051–1055 (2012). doi:10.1109/LGRS.2012.2188881

13. Gamba, P., Dell'Acqua, F., Dasarathy, B.V.: Urban remote sensing using multiple data sets: past, present, and future. Inf. Fusion **6**(4), 319–326 (2005). doi:10.1016/j.inffus.2005.02.007
14. Getoor, L., Taskar, B.: Introduction to Statistical Relational Learning. MIT Press, Cambridge (2007)
15. Gomarasca, M.A.: Basics of Geomatics. Springer, Dordrecht (2009). doi:10.1007/978-1-4020-9014-1.
16. Grêt-Regamey, A., Straub, D.: Spatially explicit avalanche risk assessment linking Bayesian networks to a GIS. Nat. Hazards Earth Syst. Sci. **6**(6), 911–926 (2006). doi:10.5194/nhess-6-911-2006
17. Kampes, B.M., Hanssen, R.F., Perski, Z.: Radar interferometry with public domain tools. In: European Space Agency, (Special Publication) ESA SP, 550, pp. 59–68 (2004)
18. Koller, D., Friedman, N.: Probabilistic Graphical Models: Principles and Techniques. MIT Press, Cambridge (2009)
19. Lee, J., Schuler, D., Lang, R., Ranson, K.: K-distribution for multi-look processed polarimetric SAR imagery. In: Proceedings of IGARSS '94–1994 IEEE International Geoscience and Remote Sensing Symposium, vol. 4, pp. 2179–2181. IEEE (1994). doi:10.1109/IGARSS.1994.399685
20. Manfreda, S., Samela, C., Gioia, A., Consoli, G.G., Iacobellis, V., Giuzio, L., Cantisani, A., Sole, A.: Flood-prone areas assessment using linear binary classifiers based on flood maps obtained from 1D and 2D hydraulic models. Nat. Hazards **79**(2), 735–754 (2015). doi:10.1007/s11069-015-1869-5
21. Martinis, S., Kersten, J., Twele, A.: A fully automated TerraSAR-X based flood service. ISPRS J. Photogramm. Remote Sens. **104**, 203–212 (2015). doi:10.1016/j.isprsjprs.2014.07.014
22. Nitti, D.O., Hanssen, R.F., Refice, A., Bovenga, F., Nutricato, R.: Impact of DEM-assisted coregistration on high-resolution SAR interferometry. IEEE Trans. Geosci. Remote Sens. **49**(3), 1127–1143 (2011). doi:10.1109/TGRS.2010.2074204
23. Papakosta, P., Straub, D.: A Bayesian network approach to assessing wildfire consequences. In: Proceedings ICOSSAR 2013, Straub 2005, pp. 3131–3138 (2013)
24. Piccarreta, M., Pasini, A., Capolongo, D., Lazzari, M.: Changes in daily precipitation extremes in the Mediterranean from 1951 to 2010: The Basilicata region, Southern Italy. Int. J. Climatol. **33**(15), 3229–3248 (2013). doi:10.1002/joc.3670
25. Pierdicca, N., Pulvirenti, L., Chini, M., Guerriero, L., Candela, L.: Observing floods from space: experience gained from COSMO-SkyMed observations. Acta Astronaut. **84**, 122–133 (2013). doi:10.1016/j.actaastro.2012.10.034
26. Pulvirenti, L., Chini, M., Pierdicca, N., Guerriero, L., Ferrazzoli, P.: Flood monitoring using multi-temporal COSMO-SkyMed data: image segmentation and signature interpretation. Remote Sens. Environ. **115**(4), 990–1002 (2011). doi:10.1016/j.rse.2010.12.002
27. Qin, D., Jianwen, M., Yun, O.Y.: Remote sensing data change detection based on the CI test of Bayesian networks. Comput. Geosci. **32**(2), 195–202 (2006). doi:10.1016/j.cageo.2005.06.012
28. RapidEye, A.G.: Satellite imagery product specifications. Tech. Rep. (2015). http://www.flyby.it/images/brochure/rapideye/eng/re_product_specifications_eng.pdf
29. Refice, A., Belmonte, A., Bovenga, F., Pasquariello, G., Nutricato, R.: On the interpolation of sparse-grid InSAR data without need of phase unwrapping. Eur. J. Remote Sens. **46**(1), 807–821 (2013). doi:10.5721/EuJRS20134648
30. Refice, A., Capolongo, D., Pasquariello, G., D'Addabbo, A., Bovenga, F., Nutricato, R., Lovergine, F.P., Pietranera, L.: SAR and InSAR for flood monitoring: examples with COSMO-SkyMed data. IEEE J. Sel. Top. Appl. Earth Obs. Remote Sens. **7**(7), 2711–2722 (2014). doi:10.1109/JSTARS.2014.2305165
31. Rignot, E.J.M., van Zyl, J.J.: Change detection techniques for ERS-1 SAR Data. IEEE Trans. Geosci. Remote Sens. **31**(4), 896–906 (1993). doi:10.1109/36.239913
32. Samela, C., Troy, T.J., Manfreda, S.: Geomorphic classifiers for flood-prone areas delineation for data-scarce environments. Adv. Water Resour. **102**, 13–28 (2017). doi:10.1016/j.advwatres.2017.01.007
33. Straub, D.: Natural hazards risk assessment using Bayesian networks. In: ICOSSAR, pp. 2509–2516

34. Touzi, R., Lopes, A.: Statistics of the stokes parameters and of the complex coherence parameters in one-look and multilook speckle fields. IEEE Trans. Geosci. Remote Sens. **34**(2), 519–531 (1996). doi:10.1109/36.485128
35. Vogel, K., Riggelsen, C., Korup, O., Scherbaum, F.: Bayesian network learning for natural hazard analyses. Nat. Hazards Earth Syst. Sci. **14**(9), 2605–2626 (2014). doi:10.5194/nhess-14-2605-2014
36. Zebker, H.A., Villasenor, J.: Decorrelation in interferometric radar echoes. IEEE Trans. Geosci. Remote Sens. **30**(5), 950–959 (1992). doi:10.1109/36.175330

Erratum to: River Flood Forecasting System: An Interdisciplinary Approach

Viacheslav Zelentsov, Ilya Pimanov, Semen Potryasaev, Boris Sokolov, Sergey Cherkas, Andrey Alabyan, Vitaly Belikov, and Inna Krylenko

Erratum to:
Chapter 4 in: A. Refice et al. (eds.), *Flood Monitoring through Remote Sensing*, Springer Remote Sensing/Photogrammetry, https://doi.org/10.1007/978-3-319-63959-8_4

The original version of this chapter was inadvertently published with incorrect spelling of one of the author "Viktor Belikov". The name has been updated as "Vitaly Belikov".

The updated online version of the original chapter can be found under https://doi.org/10.1007/978-3-319-63959-8_4

A. Refice et al. (eds.), *Flood Monitoring through Remote Sensing*, Springer Remote Sensing/Photogrammetry, https://doi.org/10.1007/978-3-319-63959-8_9

Index

A
Adaptive image processing, 118
Automatic flood detection techniques, 49, 89, 183

B
Bayesian networks (BNs), 13, 18, 154, 182–191, 194, 198, 200–206
Binary classifiers, 17, 63, 67–70

C
Change detection, 12, 14, 27–53, 117, 147, 150, 152–153, 181, 182
Color composite image, 122, 170

D
Data fusion, 13, 118, 119, 122, 181–206
DEM-based approaches, 61–76
DEM-based flood area assessment, 157, 182

F
Fast-ready flood map, 117–119, 121, 122, 126, 128, 132
Flood hazard maps, 66, 70, 71
Flood monitoring, 1–18, 29, 84, 85, 91, 101–112, 115–132, 181–206
Flood response, 27–54, 83

G
Geomorphic changes, 37, 46, 49, 50, 52, 53
Geomorphological mapping, 35, 67, 68, 109

I
Intelligent interface, 87–89, 93
Interdisciplinary approach, 81–97

M
Multimodel simulation, 85
Multi-temporal analysis, 41, 42, 49, 119
Multi-temporal flood monitoring, 2, 119, 195

O
Operational forecasting, 85, 89
Optical and SAR remotely sensed data, 49
Optical remote sensing, 4, 27–30, 39, 41–42, 46, 50, 54

R
Remote sensing data, 2, 13, 15, 16, 18, 30–32, 34, 41, 48, 85, 86, 91–92, 95, 96, 186
River floods, 27, 81–97, 101–112

S
SAR backscatter modeling, 136, 143, 149–150, 151, 154–156, 190, 198, 200
SAR-based flood monitoring, 162
SAR image segmentation, 117, 118
SAR-optical data integration, 103, 203–205
SAR polarimetry and interferometry for flood mapping, 144–147
Service oriented architecture (SOA), 85, 96

T
Time series analysis, 12, 151

A. Refice et al. (eds.), *Flood Monitoring through Remote Sensing*,
Springer Remote Sensing/Photogrammetry,
https://doi.org/10.1007/978-3-319-63959-8

FSC
www.fsc.org
MIX
Papier | Fördert
gute Waldnutzung
FSC® C083411

Kolibri 360 GmbH
Ferdinand-Jühlke-Straße 7
99095 Erfurt, Deutschland
produktsicherheit@kolibri360.de